Daniel Wilson

Brain-Weight and Size in Relation to Relative Capacity of Races

Daniel Wilson

Brain-Weight and Size in Relation to Relative Capacity of Races

ISBN/EAN: 9783337350703

Printed in Europe, USA, Canada, Australia, Japan

Cover: Foto ©berggeist007 / pixelio.de

More available books at **www.hansebooks.com**

BRAIN-WEIGHT AND SIZE

IN RELATION TO

RELATIVE CAPACITY OF RACES.

BY

DANIEL WILSON, LL.D., F.R.S.E.

PROFESSOR OF HISTORY AND ENGLISH LITERATURE, UNIVERSITY COLLEGE.

READ BEFORE THE AMERICAN ASSOCIATION FOR THE ADVANCEMENT OF SCIENCE, AT BUFFALO, N.Y., 25TH AUGUST, 1876.

[From the "Canadian Journal," October, 1876.]

TORONTO:

COPP, CLARK & CO., 67 & 69 COLBORNE STREET.

1876

BRAIN: WEIGHT AND SIZE

TO

JOSEPH BARNARD DAVIS, M.D., F.R.S., F.S.A.

THIS LITTLE BROCHURE

IS DEDICATED, IN GRATEFUL RECOGNITION OF HIS VALUABLE CONTRIBUTIONS

TO ANTHROPOLOGICAL SCIENCE.

BRAIN-WEIGHT AND SIZE

IN RELATION TO

RELATIVE CAPACITY OF RACES.

BY DANIEL WILSON, LL.D., F.R.S.E.

Read before the American Association for the Advancement of Science, at Buffalo, N.Y. 25th August, 1876.

Consistently with the recognition of the brain as the organ of intellectual activity, it seems not unnatural to assume for man, as the rational animal, a very distinctive cerebral development. One of the most distinguished of living naturalists, Professor Owen, has even made this organ the basis of a system of classification, by means of which he separates man into a sub-class distinct from all other mammalia. But while a comparison between man and the anthropoid apes, as the animals most nearly approximating to him in physical structure, lends confirmation to the idea not only that a well developed brain is essential to mental activity, but that there is a close relation between the development of the brain and the manifestation of intellectual power: the distinctive features in the human brain, as compared with those of the anthropomorpha, prove to be greatly less than had been assumed under imperfect knowledge. The substantial difference is in volume. "No one, I presume," says Darwin, "doubts that the large size of the brain in man, relatively to his body, in comparison to that of the gorilla or orang, is closely connected with his higher mental powers;" * and it might not unfairly be reasoned from analogy, that the same test distinguishes the intellectual man from the stolid, and the civilised man from the savage. A careful study of the subject, however, shows some remarkable deviations

---

* "The Descent of Man," Part I., chap. iv.

from such a scale of progression. In this Mr. Darwin would recognize an analogy to greatly more ample proofs of inequality between the organic source of power and the manifestations of mental energy; as, for example, in the ant, with its cerebral ganglia not so large as the quarter of a small pin's head, displaying instincts and apparent affections of wonderful intensity and compass. Viewed in this aspect, " the brain of an ant is one of the most marvellous atoms of matter in the world, perhaps more marvellous than the brain of man." Here, however, we look on elements of contrast rather than analogy ; and seek in vain in this direction for any appreciable test of the soundness of the popular belief in the size of the brain as a measure of intellectual power. It is otherwise when we turn to the anthropomorpha. There, alike in the scientific and in the popular creed, very special and exceptional affinities to man are admitted; and the more careful study of their anatomical structure tends to increase the recognized points of analogy.

Mr. Lockhart Clarke, in a contribution to Dr. Maudsley's work on the Physiology and Pathology of Mind, gives a minute description of the concentric layers of nervous substance which combine to form the convolutions of the human brain, and of the forms and disposition of the various nerve-cells of which its vesicular structure consists. Comparing the human brain with those of other animals, he says : " Between the cells of the convolutions in man and those of the ape tribe I could not perceive any difference whatever ; but they certainly differ in some respects from those of the larger mammalia— from those, for instance, of the ox, sheep, or cat."* Apart from the difference in volume (55 to 115 cub. in.), the only distinctive features, according to Professor Huxley, between the brain of the anthropomorpha and that of man, are " the filling up of the occipito-temporal fissure ; the greater complexity and less symmetry of the other sulci and gyri; the less excavation of the orbital face of the frontal lobe ; and the larger size of the cerebral hemispheres, as compared with the cerebellum and the cerebral nerves."

The brain of the orang is the one which seems most nearly to approximate to that of man. In volume it is about twenty-six or twenty-seven cubic inches ; or about half the minimum size of a normal human brain. The frontal height is greater than in that of

---

* " Insanity and its Treatment," by G. F. Blandford, M.D , p. 10.

other anthropomorpha; the frontal lobe is in all respects larger as compared with the occipital lobe; and certain folds of brain-sub-stance, styled "bridging convulsions," which in the human brain are interposed between the parietal and occipital lobes, also occur, though greatly reduced, in the brain of the orang; while they appear to be wholly wanting in the chimpanzee, the gibbon, and other apes which superficially present a greater resemblance to man. Referring to the convolutions of the central cerebral lobe, Huschke says: "With their formation in the ape, the brain enters the last stage of development until it arrives at its perfection in man;" and the higher class of brains may be arranged between the extremes of poorly and richly convoluted examples.

But it must not be overlooked that, apart from structural differ-ences, relative, and not absolute mass and weight of brain has to be considered, otherwise the elephant and the whale would take the fore-most place. "The brain of the porpoise," Professor Huxley remarks,* "is quite wonderful for its mass, and for the development of the cerebral convolutions;" but it is the centre of a nervous system of corresponding capacity, while as compared with the size of the animal, the brain is not relatively large. Vogt states the weight of the human body to be to the brain, on an average, as 36 to 1; whereas in the most intelligent animals the difference is rarely less than 100 to 1.

Assuming the existence of some uniform relation between the size of the brain and the development of the intellectual faculties, along with whatever is recognized as most closely analogous to them in the lower animals, it might be anticipated that we should find not only a graduated development of brain in the anthropomorpha as they approximate in resemblance to man; but, still more, that the pro-gressive stages from the lowest savage condition to that of the most civilized nations should be traceable in a comparative size and weight of brain. Dr. Carl Vogt, after discussing certain minor and doubtful exceptions, thus proceeds: "We find that there is an almost regular series in the cranial capacity of such nations and races as, since historic times, have taken no part in civilization. Australians, Hot-tentots, and Polynesians, nations in the lowest state of barbarism, commence the series; and no one can deny that the place they

---

* "Mr. Darwin's Critics: Critiques and Addresses."

occupy in relation to cranial capacity and cerebral weight corresponds
with the degree of their intellectual capacity and civilization."* But
the position thus confidently assigned to the Polynesians receives no
confirmation from the evidence supplied by the measurements of Dr.
J. B. Davis, in his *Thesaurus Craniorum;* and a careful study of
the subject reveals other remarkable deviations from such a scale of
progression, not only in individuals but in races. To these exceptional deviations, with their bearing on the comparative capacity of
races, the following remarks are chiefly directed. The largest and
heaviest brains do indeed appear, for the most part, to pertain to the
nations highest in civilization, and to the most intelligent of their
number. But this cannot be asserted as a uniform law, either in
relation to races or individuals. The more carefully the requisite
evidence is accumulated, the less does it appear that the volume of
brain, or the cubic contents of the skull, supply any uniform gauge
of intellectual capacity. In the researches which have thus far been
instituted into the characteristics of the human brain among the
lowest races, the development is in many respects remarkable; and,
as was to be expected, no organic differences between diverse races
of men have been traced.

Professor C. Luigi Calori has published the results of a careful
examination of the brain of a Negro of Guinea. It presented the
marked excess of length over breadth so characteristic of the Negro
cranium; but in other respects it corresponded generally to the fully
developed European brain. The distribution of the white and gray
substances was the same; the cerebral convolutions were collected
into an equal number of lobes; and the only special difference was
that the convolutions were a little less frequently folded, and the
separating sulci somewhat less marked than in the average European
brain. But even in this respect the complication was great. The
actual weight of the brain, according to Professor Calori, was 1,260
grammes, equivalent to 44·4 cubic inches. The complexity of convolution, and consequent extension of superficies of the encephalon, appears to be an essential element in the development of the brain as the
organ of highest mental capacity; and to the cerebrum, apparently,
the true functions of intellectual activity pertain. Professor Wagner
undertook the measurement of the convex surface of the frontal
lobe in a series of brains. The heaviest, as a rule, had also the greatest

---

* Vogt, "Lectures on Man." Lect. III.

development of surface. But the two elements were not in uniform ratio. Some of the lighter brains presented a much greater degree of convolution and consequent extent of convex superficies than others which ranked above them in weight. It is thus apparent that in estimating the comparative characteristics of brains, various elements are necessary for an exhaustive comparison. Besides the functional differences of the cerebrum, cerebellum, and pons varolii, they have different specific gravities, so that brains of equal weight may differ widely in quality. Dr. Peacock, taking distilled water as 1000, gives the values of the subdivisions of the brain thus: cerebrum, 1034; cerebellum, 1041; pons varolii, 1040. Again, Dr. Sankey states the mean specific gravity of the gray matter of the brain in either sex as 1·0346, and of the white matter as 1·0412. The variations from these results, as given by Bastian, Thurnam, and others, are trifling.' But it is significant to note that recent researches shew that where greater specific gravity of brain occurs in the insane, it appears to be limited to the gray matter.* Professor Goodsir maintained that symmetry of brain has more to do with the higher faculties than bulk or form. It is, at any rate, apparent that two brains of equal weight may differ widely in quality.

Nevertheless, the popular estimate embodied in such expressions as "a good head," "a long-headed fellow," and "a poor head," like many other popular inductions, has truth for its basis. Up to a certain stage the growth of the brain determines the capacity of the skull. Then it seems as though more complex convolutions accompanied the packing of the elaborated cerebral mass within the fixed . limits of its osseous chamber.

A comparison of races, based on minute investigation of an adequate number of brains of fair typical examples, may be expected to yield important results; but in the absence of such direct evidence, the chief data available for this purpose are derived from measurements of the internal capacity of their skulls. Among English observers who have devoted themselves to this class of observations, the foremost place is due to Dr. J. Barnard Davis, who, in 1867, summed up the results of his extensive researches in a contribution to the Royal Society, entitled, "Contributions towards determining the weight of the brain in different races of man."† Inferior as such

---

* "Journal of Mental Science," Vol. XII., p. 23.
† "Philosophical Transactions," Vol. CLVIII., p. 505.」

evidence must necessarily be, if compared with the examination of
the brain itself, nevertheless the number of skulls of the different
races gauged unquestionably furnishes some highly valuable data for
ethnical comparison. The evidence, moreover, is obtained from a
source in some respects less variable than the encephalon ; and will
always constitute a corrective element in estimating results based on
direct examinations of the brain. Dr. Davis, indeed, claims "that
the examination of a large series of skulls in ascertaining their
capacities and deducing from those capacities the average volume of
the brain, affords in some respects more available data for determining
this relative volume for any particular race than the weighing of the
brain itself." The defect is, that its most important results are
necessarily based on the assumption of a uniform density of brain ;
whereas some notable ethnical differences, hereafter referred to, may
prove to be due to the fact that certain races derive their special
characteristics from a prevailing diversity in this very respect.

But the extensive observations of Dr. Davis, as of Dr. Morton,
have a special value from the fact that each furnishes results based on
a uniform system of observation ; for the diverse methods and materials employed by different observers in gauging the human skull
have greatly detracted from their practical value. In a communi-
cation by the late Professor Jeffreys Wyman to the Boston Natural
History Society,[*] he presented the results of a series of measurements of the internal capacity of the same skull with pease, beans,
rice, flax-seed, shot, and coarse and fine sand. From repeated experiments he arrived at the conclusion that the apparent capacity varied
according to the different substances used, so that the same skull
measured respectively, with pease 1193 centimetres, with shot 1201·8,
with rice 1220·2, and with fine sand 1313 centimetres. Professor
Wyman was led to the conclusion that, for exactness, small shot, as
employed latterly by Dr. Morton, is preferable to sand, were it not
for its weight, which, in the case of old and fragile skulls, is apt to be
destructive to them. With a view to avoid the latter evil, Dr. J.
B. Davis has used fine Calais sand of 1·425 specific gravity. The
diversity in apparent volume, consequent on the employment of different substances in gauging the internal capacity of the skull, necessarily detracts from the value of comparative results of Morton, Davis,
and others. But the elaborate measurements of their great collections

of human crania furnish reliable series of data, each uniform in system, and sufficiently minute to satisfy many requirements of comparative craniometry.

Without assuming an invariable correspondence in cubical capacity and brain-weight, there is a sufficient approximation in the cubical capacity of the skull and the average weight of the encephalon to render the deductions derived from gauging the capacities of skulls of different races an important addition to this department of comparative ethnology. For minute cerebral comparisons, however, it is apparent that much more is required; and the special functions assigned to the various organs within the cranium have to be kept in view. Of these the medulla oblongata, in direct contact with the spinal cord, is now recognized as the centre of the vital actions in breathing and swallowing; and is believed also to be the direct source of the muscular action employed in speech. Next to it are the sensory ganglia, arranged in pairs along the base of the brain. To the cerebellum, which the phrenologist sets apart as the source of the emotions and passions embraced in his terminology of amativeness, philoprogenitiveness, &c., physiologists now assign the function of conveying to the mind the conditions of tension and relaxation of the muscles, and so controlling their voluntary action. But above all those is the cerebrum, or brain-proper, consisting of two large lobes of nervous substance, which in man are so large that, when viewed vertically, they cover and conceal the cerebellum. To this organ is specially assigned emotion, volition, and ratiocination. It is the assumed seat of the mind; and, in a truer sense than the skull,

"The dome of thought, the palace of the soul;"

if indeed it be not, to one class of reasoners, the mind itself. Certain it is that no acute disease can affect it without a corresponding disorder of the functions of mind; and with this organ much below the average size, intellectual weakness may always be predicated. But at the same time, it is significant to note that the human brain, stinted in its full proportions, and reduced to a seeming equality with the anthropomorpha, exhibits no corresponding capacities or instincts in lieu of the higher mental qualities. Microcephaly is the invariable index, not of mere limited intelligence and mental capacity, but of actual mental imbecility. If the augmentation of the brain of the anthropomorpha from 55 to 115 cubic inches be all that is requisite for the transformation of the irrational ape into the reasoning man, it would seem to be in no degree illogical to look for the accompaniment of the inversion of the process by an approximation, in some instances, to certain capacities and functions of the ape. But there are no indications of this. In some examples of microcephaly, the so-called animal propensities do indeed manifest themselves to excess; but there is no reproduction of the animal nature, instincts, or capacities, analogous to the scale of cerebral development of the orang or chimpanzee. A microcephalous idiot, who died at the age of twenty-two, in St. Bartholomew's Hospital, London, had a brain weighing only 13·125 oz., or 372 grammes. In describing this case, Professor Owen remarks : "Here nature may be said to have performed for us the experiment of arresting the development of the brain almost exactly at the size which it attains in the chimpanzee, and where the intellectual faculties were scarcely more developed. Yet no anatomist would hesitate in at once referring the cranium to the human species." And so is it with the encephalon. The brain of the chimpanzee is a healthy, well-developed organ, adequate to the amplest requirements of the animal; whereas the microcephalous human brain is inadequate for any efficient, continuous cerebral activity : not merely limited in its range of powers. Much, however, may yet be learned from a careful attention to the imperfect manifestations of activity in certain directions, in cases of microcephalic idiocy, and noting the predominant tendency in each case, with a view to subsequent examination of the brain. By this means it may be found possible to refer certain forms of mental activity to special variations in the structure of the organ, or to distinct members of the encephalon.

Dr. Laennec recently exhibited to the Anthropological Society of Paris a microcephalous idiot of the male sex, aged fourteen years. "This child is entirely unconscious of his own actions, and his intellectual operations are very few in number, and very rudimentary. His language consists of two syllables, *oui* and *la*, and he takes an evident pleasure in pronouncing them. He takes no heed in what direction he walks. He would step off a precipice, or into a fire." Attention was specially directed to the idiot's hands : "The thumbs are atrophied, and cannot be opposed to the other fingers. The palms of the hands have the transverse creases, but not the diagonal —the result of the atrophy of the thumbs. Hence the hand resembles that of the chimpanzee. The dentition too is defective. Though fourteen years of age, the child has only twelve teeth." Here it is curious to note the analogies in physical structure to the lower anthropomorph in other organs besides the brain, for it only renders more striking the absence of any corresponding aptitudes.

Dr. J. Barnard Davis, in his interesting monograph on "Synostotic Crania among Aboriginal Races of Man," produces some remarkable illustrations of the effect of premature ossification of the sutures of the skull in arresting the full development of the brain, and so rendering it unequal to the due performance of its functions. "I have," he says, "the cranium of a convict who was executed on Norfolk Island, which I owe to the kindness of Admiral H. M. Denham. This man was executed there when that beautiful isle was appropriated to the reception of the most dangerous and irreclaimable convicts from the other penal settlements. It is a microcephalic skull, rather dolichocephalic, of a man apparently about forty years of age. It exhibits a perfect ossification of the sagittal and of the greater portion of the lambdoidal sutures. The coronal suture is partially obliterated at the sides in the temporal regions, and can only be distinguished by faint traces in all its middle parts. In this case there has not been any compensatory development of moment in other directions. The calvarium is not abridged in its length, which is 7·1 inches, equal to 179 millimetres; probably it is a little elongated. It is, however, very narrow, being only 4·8 inches, or 122 m.m. at its widest part, between the temporal bones. So that the result is a very small, dwarfed, almost cylindrical calvarium. The internal capacity is only 59 ounces of sand,* which is equal to 71·4 cubic inches, or 1169 cubic centimetres." Here is a skull considerably below the lowest mean of the crania of any race in Morton's enlarged tables, or in the more comprehensive ones furnished in Dr. Davis's "Thesaurus Craniorum." Another skull nearly approximating to it is that of a Cole, one of the savage tribes of Nagpore, in Central India, who are said to go entirely naked. It is described in the supplement to the "Thesaurus Craniorum" as that of "Chara," a Cole farmer, aged fifty,

---

* The internal capacity of 59 oz. is given here from the "Thesaurus Craniorum," p. 40, in correction of that of 50 oz. stated in the memoir in "Transactions of the Dutch Society of Sciences," Haarlem, p. 21, which may be presumed to be a misprint. Dr. Davis adds, in the "Thesaurus Craniorum," "An early closure of the sutures has occasioned a stunted growth of the brain, especially of its convolutions, and thus prevented the development of those structures and faculties which might have given a different direction to his lower propensities;" and he justly adds his conviction that this was a case rather for timely treatment as a dangerous idiot, than for punishment as a criminal.

and its internal capacity is stated as 59·5 oz. av., equivalent to 71·7 cub. inches. The Coles appear to be small of stature. The heights of three of them, whose skulls are in the same collection, were respectively 5 ft. 5 in., 5 ft. 2 in., and 5 ft., and the average internal capacity of five male skulls is only 66·6. The small stature in this and others of the native races of Central India, has to be taken into account in estimating the relative size of the brain. But the Cole skulls are remarkable for their small size, being smaller even than the ordinary Hindoos of Bengal. Yet one of them, "Cootlo," whose skull is among those included in the above mean, commanded a band of insurgents in the Porahaut rebellion of 1858, and made himself a terror to the district.

The microcephalism of races, as well as of individuals, of small stature, must not be confounded with the true microcephaly of a dwarfed or imperfectly developed brain, which is invariably accompanied with mental imbecility. The Mincopies of the Andaman Islands are spoken of by Professor Owen as "perhaps the most primitive, or lowest in the scale of civilization, of the human race."[*] Mr. G. E. Dobson, in describing his first visit to one of their "homes," says : "Although none of the tribe exceeded 64 inches in height, so that on first seeing them we thought the shed contained none but boys and girls, I was especially struck by the remarkable contrast between the size of the males and females."[†] Dr. J. B. Davis has given, in the supplement to "Thesaurus Craniorum," the dimensions of a male Mincopie skeleton in his collection. The age he assumes to have been about thirty-five. The internal capacity of the skull is 62 oz. (Calais sand), equivalent to 75·5 cubic inches, and the entire height of the skeleton is 58·7 inches. It belongs, says Dr. Davis, to a pigmy race, is small in all its dimensions, and is particularly small in the dimensions of the pelvis. Of their skulls, moreover, he adds, "it is somewhat difficult to determine the sex with confidence. They are all small (but this is a character of the race), they are delicate in development, and they have that fullness of the occipital region, and smallness of the mastoid processes, which are marks of feminism."

Mr. Alfred R. Wallace connects the Mincopies with the Negritos and Semangs of the Malay peninsula, a dark woolly-haired race, dwarfs in stature. Dr. Davis says of the six Mincopie skulls in his collection, four male and two female, as well as of others which he has seen : "They are all remarkably and strikingly alike, not merely in size but in form also. They are all small, round, brachycephalic crania of beautiful form." Moreover, though classed as "lowest in the scale of civilization," the Mincopies betray no deficiency of intellect. The admirable photographs which illustrate Mr. Dobson's narrative show in the majority of them good frontal development. The brain is not, indeed, relatively small. Their canoes are made of the trunk of a tree, hollowed out; and Mr. Dobson remarks : "The construction of their peculiar arrows and fish spears with movable heads exhibits much ingenuity, and the use of no small reasoning power in adapting means to an end."

We are indeed too apt to apply our own artificial standards as the sole test of intellectual vigour; whereas it is probable that in the amount of acquired knowledge and acuteness of reasoning many savage races surpass the majority of the illiterate peasantry in the most civilized countries of Europe. Mr. Wallace, in viewing the subject in one special light, remarks : "The brain of the lowest savages, and, as far as we yet know, of the prehistoric races, is little inferior in size to that of the higher types of man, and is immensely superior to that of the higher animals; while it is universally admitted that quantity of brain is one of the most important, and probably the most essential of the elements which determine mental power. Yet the mental requirements of savages, and the faculties actually exercised by them are very little above those of animals. The higher feelings of pure morality and refined emotion, and the power of abstract reasoning and ideal conception, are useless to them; are rarely, if ever, manifested; and have no important relations to their habits, wants, desires, and well-being. They possess a mental organ beyond their needs."*

Here, however, it may be well to guard against the confusion of two very distinct elements. The higher feelings of pure morality and refined emotion are not manifestations of intellectual vigour in the same sense as is the power of abstract reasoning and ideal conception. It is not rare to find an English or Scottish peasant with little intellectual culture or capacity for abstract reasoning, but with an acutely instinctive moral sense. On the other hand, among the criminal

---

* "Limits of Natural Selection, as Applied to Man."

class, it is by no means rare to find examples of wonderfully vigorous intellectual power applied to the planning and accomplishing of schemes which involve as much foresight and skill as many a triumph of diplomacy; but which at the same time seem to be nearly incompatible with any moral sense. Moreover, it is needless to say that intellectual vigour and high moral principle are by no means invariable concomitants in any class of society; nor can they be traced to a common source. Mr. Wallace recognizes that "a superior intelligence has guided the development of man in a definite direction, and for a special purpose;" and such guidance involves much more than the mere evolution of a higher animal organization. But, appreciating as he does the difficulties involved in any acceptance of a theory of evolution which assumes man to be the mere latest outgrowth of a development from lower forms of animal life, Mr. Wallace points out that "natural selection could only have endowed savage man with a brain a little superior to that of an ape, whereas he actually possesses one very little inferior to that of a philosopher."

Yet neither Mr. Wallace, nor Professor Huxley when controverting this argument, withholds a due recognition of the activity of the intellect of the savage. No one indeed can have much intercourse with savage races wholly dependent on their own resources, without recognizing that, within a certain range, their faculties are kept in constant activity. The savage hunter has not merely an intimate familiarity with all the capabilities and resources of many regions traversed by him in pursuit of his game; his geographical information includes much useful knowledge of the topography of ranges of country which he has never visited. I found, on one occasion, when exploring the Nepigon River, on Lake Superior, that my Chippewa guides, though fully five hundred miles from their own country, and visiting the region for the first time, were nevertheless on the look-out for a metamorphic rock underlying the sienite which abounds there; and they made their way by well-recognized landmarks to this favourite "pipe-stone rock." While moreover the Indian, like other savages, is devoid of much of what we style "useful knowledge," but which would be very useless to him, he is fully informed on many subjects embraced within the range of the natural sciences; and has a very practical knowledge of meteorology, zoology, botany, and much else which constitutes useful knowledge to him. He is familiar with the habits of animals, and the medicinal virtues of many plants; will find his way through the forest by noting the special side of the trunks on which certain lichens grow; and follow the tracks of his game, or discover the nests of birds, by indications which would escape the most observant naturalist. The Australian savage, stimulated apparently to an unwonted ingenuity by the privations of an arid climate, is the inventor of two wonderfully ingenious implements, the *wommera* or throwing stick, and the *bomerang*, which, when employed by the native expert, accomplish feats entirely beyond any efforts of European skill. Moreover, as Professor Huxley remarks, he "can make excellent baskets and nets, and neatly fitted and beautifully balanced spears; he learns to use these so as to be able to transfix a quartern loaf at sixty yards; and very often, as in the case of the American Indians, the language of a savage exhibits complexities which a well-trained European finds it difficult to master." Again he goes on to say: "Consider that every time a savage tracks his game he employs a minuteness of observation, and an accuracy of inductive and deductive reasoning which, applied to other matters, would assure some reputation to a man of science, and I think we need ask no further why he possesses such a fair supply of brains. In complexity and difficulty, I should say that the intellectual labour of a good hunter or warrior considerably exceeds that of an ordinary Englishman." Hence Professor Huxley is not prepared to admit that the American or Australian savage possesses in his brain a mental organ which he fails to turn to full account. But without entering on the questions of evolution and natural selection in all their comprehensive bearings, it is still apparent that the brain of the savage is an instrument of great capacity, employed within narrow limits.

In estimating the comparative size of the brain, it is seen to be necessary to discriminate between individuals or races of small stature and cases of true microcephaly. On the other hand, it is not to be overlooked that examples of idiocy are not rare where the head is of a fair average size, and where the mental imbecility is regarded as congenital. But in this as in other researches of the physiologist, he is limited in his observations mainly to the chance opportunities which offer for study; and not unfrequently the prejudices of affection arrest the hand of the student, and prevent a *post mortem* examination

in cases where science has much to hope for from freedom of investigation. Hence the data thus far accumulated in evidence of the actual
structure, size and weight of the human brain fall far short of what is
requisite for a solution of many questions in reference to the relations
between cerebration and mental activity.   From time to time men of
science have sought by example, as well as by precept, to lessen such
impediments to scientific research.   Dr. Dalton left instructions for
a *post mortem* examination, in order to test the peculiarity of his
vision, which he had assumed to be due to a colouring of the vitreous
humour ; Jeremy Bentham bequeathed his body to his friend Dr.
Southwood Smith, for the purposes of anatomical science ; and the
Will of Harriet Martineau, who died during the present year, contains this provision : " It is my desire, from an interest in the progress
of scientific investigation, that my skull should be given to Henry
George Atkinson, of Upper Gloucester Place, London, and also my
brain, if my death should take place within such distance of his then
present abode as to enable him to have it for purposes of scientific
investigation." The Will is dated March 10th, 1864 ; but by a
codicil, dated October 5, 1871, this direction is revoked, with the
explanation which follows in these words : " I wish to leave it on
record that this alteration in my testamentary directions is not caused
by any change of opinion as to the importance of scientific observation on such subjects, but is made in consequence merely of a change
of circumstances in my individual case."   The natural repugnance of
surviving relatives to any mutilation of the body must always tend
to throw impediments in the way of such researches ; though it may
be anticipated that, with the increasing diffusion of knowledge, such
obstacles to its pursuit will be diminished.   Thus far, however,
notwithstanding the persevering labours of Welcker, Bergmann,
Parchappe, Broca, Boyd, Skae, Owen, Thurnam, and other physiologists, their observations have been necessarily limited almost exclu-
sively to certain exceptional sources of evidence, embracing to a large
extent only the pauper and the insane classes ; and in the case of the
latter especially, the functional disorder or chronic disease of the
organ under consideration renders it peculiarly desirable that such
results should be brought, as far as possible, into comparison with a
corresponding number of observations on healthy brains of a class
fairly representing the social and intellectual status of a civilized
community.

The average brain-weight of the human adult, as determined by a numerous series of observations, ranges for man from 40 oz. to 52½ oz., and for woman from 35 oz. to 47½ oz. But some indications among ancient crania tend to suggest a doubt as to whether this difference in cerebral capacity was a uniformly marked sexual distinction among early races; due allowance being made for difference in stature. Dr. Thurnam made the race of the British Long Barrows a special subject of study; and Dr. Rolleston has followed up his researches with valuable results. Amongst other points, he notes that the males appeared to have averaged 5 feet 6 inches, and the females 4 feet 10 inches in height. But while the difference of stature between the male and the female exceeds what is observable in most modern races, the variation in the size and internal capacity of their skulls appears to be less than among civilized races. The like characteristics are noticeable in the larger race of Europe's palæo-technic era. Nothing is more striking in the discovery of those ancient remains of European man than the remarkable development of the skulls, and the good brain capacity of the race of the palæo-technic dawn, where man is proved, by his works of art and all the traces of his hearth and home, to have been still a rude hunter and cave-dweller. Whatever other changes, therefore, may have affected the brain as the organ of human thought and reasoning, it does not thus far appear that the average mass of brain has increased since the advent of European man. Important exceptions have indeed been noted. Professor Broca's observations on the cerebral capacity of the Parisian population at different periods, based on nearly 400 skulls derived from vaults and cemeteries of various dates from the 11th or 12th to the 19th century, appear to him to show a progressive cerebral development in that remarkable centre of European civilization.[*] But though the assumption is not inconsistent with other results of civilization, and is the necessary corollary of the postulate that intellectual activity tends to permanent development of brain; the fact that the crania presented a still greater diversity in type than in size reminds us of the intermixture of races on the banks of the Seine, and the consequent necessity for much more extended observations before so important a deduction can be received as an established truth.

Taking the average brain-weight of the human adult as already stated, all male brains falling much below 40 oz. or 1130 grammes,

---

and female brains below 35 oz. or 990 grammes, may be classed as *microcephalous;* and all above the maxima of the medium male and female brain, viz., 52½ oz. or 1480 grammes, and 47½ oz. or 1345 grammes, may be ranked as *megalocephalous,* or great brains.

Professor Welcker, who devoted special attention to the whole subject under review, assumes another and simpler test, when he says that skulls of more than 540 to 550 millemetres, or 21·26 to 21·65 inches in circumference—the weight of brain belonging to which is 1490 to 1560 grammes (52·5—55 oz. avoir.),—are to be regarded as exceptionally large. But while an excess of horizontal circumference may be accepted as indicating good cerebral capacity, it must not be overlooked that the adoption of it as the key to any definite or even approximate brain-weight ignores the important elements of variation involved in the difference between acrocephalic and platycephalic head-forms. The volume of brain in Scott, and probably in Shakespeare, appears to have depended more on its elevation than its horizontal expansion. The same was also the case with Byron. The intermastoid arch, measured across the vertex of the skull from the tip of one mastoid process to the other, furnishes an accurate gauge of this development. Of thirteen selected male English skulls in Dr. Davis's collection, the mean of this measurement is 15·1; and of thirty-nine male and female English skulls, it is only 14·4. Of the whole number of eighty-one English skulls described in the "Thesaurus Craniorum," three exceptionally large ones are—No. 123, that of an ancient British chief, of fully 6 ft. 2 in. in stature, from the Grimsthorpe Barrow, Yorkshire; No. 905, a calvarium of great magnitude, very brachycephalic, and with the elevation across the middle of the parietals apparently exaggerated by compression in infancy, from Hythe, Kent; and No. 1029, another male skull, remarkable alike for its size and weight, and with a peculiarity of conformation ascribed by Dr. Davis to synostosis of the coronal suture. The intermastoid arch in those exceptionally large skulls measures respectively 16·0, 16·2 and 16·9; whereas the same measurement derived from the cast of Scott's head taken after death, yields the extraordinary dimensions of 19 inches.* This last measurement is over the hairy scalp. But after making ample allowance for this, the vertical measurement of the skull and consequently of the brain is remarkable.

---

* I am indebted to Dr. J. A. Smith F.S.A., Scot., for this and other measurements of casts of The Bruce, Burns, Scott, &c., not accessible to me.

Full value has been assigned at all periods to the well-developed forehead. It is characteristic of man. The physiognomist and the phrenologist have each given significance to it in their respective systems; and it has received no less prominent recognition from the poets. A fully developed forehead is assumed as distinctive of the male skull. But Juliet, in "The Two Gentlemen of Verona," when depreciating her rival, exclaims, "Ay, but her forehead's low;" and the jealous Queen of Egypt, in "Antony and Cleopatra," is told of Octavia that "her forehead is as low as she would wish it." "The fair large front" of Milton's perfect man is the external index of an ample cerebrum : the organ to which the seat of consciousness, intelligence, and will is assigned. It is therefore consistent with this that a low, retreating forehead is popularly assumed to be the characteristic index of the savage, and of the unintellectual among civilized races. But the cerebral characteristics of both ancient and modern civilized races have still to be studied in detail; and the influence of race and sex on the form of the head and the mass and weight of the brain, involves some curious questions in relation to the oldest illustrations of the physical characteristics of man, and to the effect of civilization on the relative development of the sexes.

Early observations led Dr. Pruner-Bey and other ethnologists of France to recognize in certain ancient Gaulish skulls of a brachycephalic type the evidences of a primitive race, assumed to represent the inhabitants of France and of Central Europe during its reindeer period, and which appeared to be assigned with reasonable probability to a Mongol origin. But in the Cro-Magnon cavern, and in other caves more recently explored, the remains of a race of men have been brought to light markedly dolichocephalic, and no less striking in cranial capacity. Dr. Broca speaks of these ancient cave-dwellers of the valley of the Vezère as characterized by "sure signs of a powerful cerebral organization. The skulls are large. Their diameters, their curves, their capacity, attain, and even surpass, our medium skulls of the present day. The forehead is wide, by no means receding, but describing a fine curve. The amplitude of the frontal tuberosities denotes a large development of the anterior cerebral lobes, which are the seat of the most noble intellectual faculties." Alongside of the remains of this ancient race, and in the underlying deposits, lay those of the mammoth, cave-lion and bear, fossil horse, and reindeer. In neighbouring caves of the same valley, and especially

2

in that of La Madelaine, numerous specimens of primitive art have been found : tools and weapons of flint, carved lances and harpoons of bone; and ingenious engravings and carvings of the mammoth, reindeer, and of man himself, on pieces of horn and ivory tablets. The evidences of primitive skill and intellectual vigour are remarkable. Dr. Broca, after a review of their ingenious arts, says : " They had advanced to the very threshold of civilization;" and Dr. Pruner-Bey thus comments on their characteristics : " If we consider that its three individuals had a cranial capacity much superior to the average at the present day ; that one of them was a female, and that female crania are generally below the average of male crania in size ; and that nevertheless the cranial capacity of the Cro-Magnon woman surpasses the average capacity of *male* skulls of to-day, we are led to regard the great size of the brain as one of the more remarkable characters of the Cro-Magnon race. This cerebral volume seems to me even to exceed that with which at the present day a stature equal to that of our cave-folks would be associated : whilst the skulls from the Belgium caves are small, not only absolutely, but even relatively in the rather small stature of the inhabitants of those caves."*

The remarkable cranial capacity of the skulls thus seemingly pertaining to the most primitive of European races—the troglodytes of the mammoth and reindeer periods of Central Europe,—is the more significant from its bearing on the evidence of progressive cerebral development adduced by Dr. Broca from skulls recovered from ancient and modern cemeteries of Paris. It appears indeed to conflict with any theory of a progressive development from the Troglodyte of the post-glacial age to the civilized Frenchman of modern times. Mr. W. Boyd Dawkins has accordingly been at some pains in his " Cave Hunting," to show that the conclusions formed by previous observers as to the epoch of their burial are not supported by the facts of the case ; and he sums up his review of the whole evidence by expressing a conviction that he " should feel inclined to assign the interments to the neolithic age, in which cave-burial was so common. The facts," he adds, " do not warrant the human skeletons being taken as proving the physique of the palæolithic hunters of the Dordogne, or as a basis for an inquiry into the ethnology of the palæolithic races." Mr. Boyd Dawkins also pronounces the same doubts in reference to the equally characteristic

---

* " Reliquiæ Aquitanicæ."

male skeleton found in a cave at Mentone, and to others obtained in the Lombrive and other caves. Nor was this caution without reason, for the remains of man differ from other animal remains found in such series of deposits as mark a succession of periods, in so far as they pertain to the only animal habitually given to the practice of interment; so that human skeletons found under such circumstances may have been artificially intruded long subsequent to the accumulation of the breccia in which they lay. Happily, however, any doubts as to the contemporaneity of the human remains with the other cave-relics has since been removed by the discovery of skeletons, similar in type, in other caverns in the same valley—and especially in that of Laugerie Basse,—in positions which seem to leave no room for questioning their being of the same age as the works of art found along with them.

Other examples of the ancient man of Europe show him in like manner endowed with a cerebral development far in advance of the rudest races of modern times. The skull found by Dr. Schmerling in the Engis Cave, near Liége, along with remains of six or seven human skeletons, was embedded in the same matrix with bones of the fossil elephant, rhinoceros, hyæna, and other extinct quadrupeds. It is a fairly proportioned, well developed dolichocephalic skull; and, like others of the seemingly most ancient human skulls yet found, has signally disappointed the expectations of those who count upon invariably finding a lower type the older the formation in which it occurs. "Assuredly," says Professor Huxley, "there is no mark of degradation about any part of its structure. It is, in fact, a fair average human skull, which might have belonged to a philosopher, or might have contained the thoughtless brain of a savage." Even the famous Neanderthal skull, of doubtful geological antiquity, but pronounced to be "the most brutal of all human skulls," acquires its exceptional character chiefly from the abnormal development of the superciliary region.

It is a universally accepted fact that the size of the male head and the weight of the brain are greater than those of the female. The average weight of the male brain is found to exceed that of the female by about ten per cent.; or, as it is stated by Professor Welcker, the brain-weight of man is to that of woman as 100 : 90. But the difference of stature between the two sexes has to be taken.

into account. The average, based on various series of observations to determine the mean stature for man and for woman, shows the latter to be about eight per cent. less than the former; or, as Dr. Thurnam has stated it more precisely:

RATIO OF STATURE AND BRAIN-WEIGHT IN THE TWO SEXES.

	MALE.		FEMALE.
Stature	100·	......	92·
Weight of Brain	100·	......	90·3

Here again, however, it becomes important to take into consideration other elements of difference besides weight; for, as Tennyson insists, "Woman is not undevelopt man, but diverse." The results of Wagner's observations on the superficial measurements of the convolutions of the brain point to the conclusion that in the female the lesser brain-weight may be compensated by a larger superficies. Ranked in the order of their relative weights in grammes, six average brains of men and women were found to stand thus:

1	Male	(a)	1340
2	"	(b)	1330
3	"	(c)	1273
4	Female	(d)	1254
5	"	(e)	1223
6	"	(f)	1185

But the same brains, when tested by the degrees of convolution of the frontal lobe, measured in squares of sixteen square millimetres, irrespective of the question of relative size, ranked as follows, advancing the female (d) from the fourth to the first place, and reducing the male (c) from the third to the sixth place:

1	Female	(d)	2498
2	Male	(a)	2451
3	Male	(b)	2309
4	Female	(f)	2300
5	Female	(e)	2272
6	Male	(c)	2117

But, as already indicated, some modern disclosures tend to raise the question whether the difference between the sexes, in so far as relative volume of brain is concerned, has not been increased as a result of civilization. The disparity in size between the Cro-Magnon male and female skeletons is quite as great as that of modern times, but the capacity of the female skull is relatively good; and M. Broca, in his paper on the "Caverne de l'Homme Mort," says: "One of the most remarkable traits of the series is the great relative capacity of the female skulls."

Other observations, such as those of Professor Rolleston " On the People of the Long Barrow Period," seem to indicate a nearer approximation in actual cranial capacity of the two sexes in prehistoric times than among modern civilized races. On the assumption that intellectual activity tends to permanent development of brain, it is consistent with the conditions of savage life that it should bring the mental energies of both sexes into nearly equal play. They have equally to encounter the struggle for existence, and have their faculties stimulated in a corresponding degree. As nations rise above the purely savage condition of the hunter stage, this relative co-operation of the sexes is subjected to great variations. The laws of Solon with reference to the right of sale of a daughter or sister, and the penalties for the violation of a free woman, show the position of the weaker sex among the Greeks at that early stage to have been a degrading one. But the change was great at a later stage; and much of our higher civilization is traceable to the early establishment of the European woman's rights, which Christianity subsequently tended to enlarge. The position of woman among the ancient Britons appears to have been one of perfect equality with man. Among the Arabians and other Mohammedan nations, including the modern Turks, the opposite is the case; and the whole tendency of the creed of the Koran, and the social life among Mohammedan nations, must be towards the intellectual atrophy of woman. Hence it is consistent with the diverse conditions of life that, in so far as cerebral development is the result of mental activity, a much closer approximation is to be looked for in the mass and weight of brain in the two sexes among savage races, than among nations where woman systematically occupies a condition of servile degradation, or of passive inertness.

Some interesting results of the actual brain-weights of Negroes and other typical representatives of inferior savage races have been published, including examples of both sexes; and although the observations are as yet too few for the deduction of any absolute or very comprehensive conclusions, they furnish a valuable contribution towards this department of ethnical comparison. In 1865, Dr. Peacock published the results of observations on the brains of four Negroes and two Negresses; and to those he subsequently added a seventh example. Other examples are included in the following table. But I have excluded some extremes of variation, such as the two given by Mascagni, one of which weighed 1458 grammes, or

51·5 oz. av., and the other only 738 grammes, or 26·1 oz. av. In addition to such actual brain-weights, Morton, Tiedemann, Davis, Wyman, and others, have gauged the skulls of Negroes, American Indians, Mincopies, Tasmanians, Australians, and other savage races, as well as those of many civilized and semi-civilized nations, and thereby contributed valuable data towards determining their relative cranial capacity. In his "Crania Ægyptiaca," Dr. Morton, when discussing the traces of a Negro element in the ancient Egyptian population, says: "I have in my possession seventy-nine crania of Negroes born in Africa, for which I am indebted to Drs. Goheen and McDowell, lately attached to the medical department of the colony of Liberia, in western Africa; and especially to Don Jose Rodriguez Cisneros, M.D., of Havana, in the island of Cuba. Of the whole number, fifty-eight are adult, or sixteen years of age and upwards, and give eighty-five cubic inches for the average size of the brain. The largest head measures ninety-nine cubic inches; the smallest but sixty-five. The latter, which is that of a middle-aged woman, is the smallest adult head that has hitherto come under my notice."*

TABLE I.

NEGRO BRAIN-WEIGHT.

	RACE.	AUTHORITY.	WEIGHT.
M	African, Mozambique........	Peacock	43·80
M	"	"	45·80
M	" Buenos Ayres........	"	44·00
M	" Congo	"	46·25
M	"	"	42·80
M	"	Sœmmering	45·40
M	"	Tiedemann	35·20
M	" Congo	C. Luigi Calori........	44·40
M	"	Barkow.................	50·80
M	"	"	45·90
M	"	"	38·90
M	"	Sir A. Cooper..........	49·00
F	Hottentot Venus..........	Marshall........	31·00
F	Bushwoman	"	30·75
F	"	"	31·50
F	"	"	31·00
F	"	Flower & Murie.........	38·00
F	African	Peacock	46·00
F	"	"	41·00

* "Crania Ægyptiaca," p. 21.

The influence of race on the volume, weight, disposition, and relative proportions of the different subdivisions of the human brain, and so of brain on the character of races, has thus far been very partially tested. But the diversities of race head-forms—brachycephalic, dolichocephalic, platycephalic, acrocephalic, &c.—are now well recognized, though their relation to cerebral development still requires much research for its elucidation. The ancient Roman forehead, as illustrated by classic busts, and confirmed by genuine Roman skulls, was low but broad, and the whole head was platycephalic. The Greek had a high forehead, and the works of the Greek sculptors show that this was regarded as typical. But contemporary with the classic races were the Macrocephali of the Euxine and the Caspian Seas, who, like many modern tribes of the New World, purposely aimed at depressing a naturally receding forehead, and thereby exaggerated the typical forehead characteristic of certain ancient barbaric races.

In the case of hybrids the interchange of physical and mental characteristics of the parents, including modifications of head-form, is a familiar fact. The English head-form appears to be an insular product of intermingled Briton, Teuton and Scandinavian elements, which has no continental analogue; and its sub-divisions, or subtypes, vary with the ethnical intermixture. The Scottish head appears to exceed the English in length, while the latter is higher. Where the Celtic element most predominates, the longer form of head is found; but even in the most Teutonic districts the difference between the prevailing head-form and that of the continental German is so marked that the latter finds it difficult to obtain an English-made hat which will fit his head.* Here the diversities of head-form are accompanied with no less marked differences of individual and national character.

Professor Welcker determined the average capacity of the German male skull as 1450 cubic centimetres, equivalent to 88 cubic inches, and representing an average brain-weight of 49 oz. Dr. Davis, by a similar process, assigns to the Germans, male and female, the larger mean brain-weight of 50·28 oz.; but by combining the means of both sexes, as derived from his own tables and those of Huschke and Wagner, we obtain a mean weight of German brain of 1314 grms. or 46·37 oz. The results of an extensive series of observations by

---

* Vide "Physical Characteristics of the Ancient and Modern Celt." *Canadian Journal*, VoL VII., p. 369.

Dr. Broca, on the male French skull, yield a mean capacity of 1502
cubic centimetres, or 91 cubic in., representing an average brain-
weight of 50·6 oz. Morton, taking his average from five English
skulls, gives the great internal capacity of 96 cubic in.; while Dr. J.
B. Davis arrives at a capacity of only 90·9 cubic in., from the
examination of thirty-two skulls, male and female; and for the
Scottish and Irish, each of 91·2 cubic in., from an examination of
thirty-five skulls. But unfortunately the Davis collection, so rich in
other respects, derived its chief English specimens from a phrenological collection ; and, along with a few large skulls, contains "many
small and poor English examples."* The average weight of the
English brain may therefore, as Dr. Davis admits, be assumed to be
higher than the mean determined by him. " Still a comparison with
actually tested weights of brains shows that there cannot be any
material error." The average brain-weight of twenty-one English-
men, as given by him, is 50·28 oz., that of thirteen women is 43·13 ;
and of the combined series, 47·50. The results determined by the
same process in relation to the other nationalities of Europe are
exhibited in detail in Dr. Davis's tables, printed in the "Philosophical
Transactions."

Such averages are, at best, only approximations to true results ;
and when obtained, as in Morton's English race, from a very few
examples, or in Dr. Davis's, from exceptional skulls, collected under
peculiar circumstances or for a special purpose, they must be tested
by other observations. According to Dr. Morton, for example, the
mean internal capacity of the English head is 96 cubic in., while that
of the Anglo-American is only 90 cubic in. Such a conclusion, if
established as the result of comparison of a sufficiently large number
of well authenticated skulls, would be of great importance in its
bearing on the influence of change of climate, diet, habits, &c., as
elements affecting varieties of the human race. But determined as
it was in the Morton collection, from five English and seven Anglo-
American specimens, it can be regarded as no more than a mere
chance result. Ranged nearly in the order of mean internal capacity
of skull, the following are the results arrived at, mainly by gauging
the skulls in various collections available for such comparisons of
different races of mankind. In presenting them here, I avail myself
of Dr. Thurnam's researches, augmenting them with other data sub-

---

* "Thesaurus Craniorum ;" App., p. 347.

sequently published, including results deduced from Dr. J. B. Davis's minute reports of his own extensive collections, and taking Tiedemann's capacity of 92·3 for the European skull as 100.

TABLE II.

RATIO OF CUBICAL CAPACITY OF SKULLS OF DIFFERENT RACES.

RACE.	AUTHORITY.	CAPACITY.
European..................	Tiedemann	100·
Asiatic....................	Davis.....................	94·3
African....................	"	93·
American..................	Tiedemann	95·
"	Davis.....................	94·7
"	Morton	87·
Oceanic..................	Davis.....................	96·9
Chinese..................	Davis.....................	99·8
Mongol...................	Morton	94·
"	Tiedemann	93·
Hindoo...................	Davis.....................	89·4
Malay	Tiedemann	89·
American Indian...........	Morton	91·
Esquimaux	Davis.	98·8
Mexican	Morton	88·5
Negro.....................	Tiedemann	91·
"	Peacock..................	88·
Hottentot.................	Morton	86·
Javan....................	Davis.....................	94·8
Tasmanian	"	88·
Australian	Morton	88·
"	Davis.....................	87·9
Peruvian.................	Wyman...................	81·2
"	Morton	81·2

The tables of Dr. Morton and Dr. Davis furnish materials for drawing comparisons between diverse nations of the great European family; but though they are of value as contributions to the required means for ethnical comparison, they fall far short of determining the average cranial capacity of the different nationalities. Whilst, for example, the tabular data in the "Thesaurus Craniorum" show a mean internal capacity of 94 cub. in. for the combined Teutonic family, the Finns yield the higher mean capacity of 96·3 cub. in. Again, Dr. Thurnam found that the results of the weighing of fifty-nine brains of patients at the Friends' Retreat near York, mostly persons of the middle class of society, yielded weights considerably above those which he subsequently obtained from testing those of pauper patients in Wilts and Somerset. But this has to be estimated along with

the undoubted ethnical differences which separate the population of
Yorkshire from that of Somerset and Wiltshire. An interesting
paper in the West Riding Asylum Reports gives the results of the
determination of 716 brain-weights, rather more than half being
males. The average is 48·149 oz. for the male, and 43·872 for the
female brain ; whereas the average weights of 267 male brains of a
similar class of patients in the Wilts' County Asylum, as given by
Dr. Thurnam, is 46·2 oz., and of 213 female brains, 41·0 oz. The
results of the observations carried on by Dr. Boyd at St. Marylebone
yield, from 680 male English brains, a mean weight of 47·1 oz., and
from 744 female brains a mean weight of 42·3 oz.; whereas Dr.
Peacock determined, from 183 cases in the Edinburgh Infirmary,
the weight of the male Scottish brain to average 49·7, and that of
the female brain to average 44·3 oz. Here the results are determined by so numerous a series that they might be accepted as
altogether reliable, were it not that in the former case they are
based to a large extent on a purely pauper class; whereas the
patients of the Royal Infirmary of Edinburgh include respectable
mechanics and others from many parts of Scotland, among whom
education is common. It is not to be doubted, indeed, that a con-
siderable difference in the form and size of the head, and no doubt
also in brain-weight, is to be looked for amongst English, Scotch,
Irish, German and French men and women, according to the county
or province of which they are natives, and the class of society to
which they belong.

The comparative ratio of the cubical capacity of the skull, or the
average brain-weight, in so far as either is indicative of ethnical differences among members of the European family of nations, has thus
to be determined by numerous examples; or dealt with in detail
in reference to the different nationalities. Even in single provinces
or counties, social position, and probably education, must be taken
into account ; so that a series of observations on hospital and pauper
patients may be expected to fall below the general average; and
fallacious comparisons between European peoples may be based on
data, correct enough *per se*, but unjust when placed alongside of a
different class of results. The great mass of evidence in reference
to brain-weight has thus far been mainly derived, in the case of the
sane, from one rank of life. A comparison of the results with those
derived from the insane of various classes of society shows less dis-

crepancy than might have been anticipated. But there are certain cases of hydrocephalous and other abnormally enlarged brains which have to be rigorously excluded from any estimate of the size or weight of the brain, either as a race-test or as an index of comparative mental power.

Were it possible to select from among the great intellects of all ages an adequate series of representative men, and ascertain their brain-weights, or even the cubical capacity of their skulls, one important step would be gained towards the determination of the relation between size of brain and power of intellect. But we have little other data than such hints as the busts of Æschylus, Pericles, Socrates, Plato, Aristotle, and other leaders of thought may supply. Malcolm Canmore—Malcolm of the great head, as his name implied,—stands forth with marked individuality from out the shadowy roll of names which figure in early Scottish history. Charlemagne, we should fancy, merited a similar designation. But the portraits of his modern imperial successor, Charles V., show no such loftiness of forehead. Judging from the portraits and busts of Chaucer, Shakespeare, Milton, Cromwell, Napoleon, and Scott, their brains must have considerably exceeded the ordinary size. In the report of the *post mortem* examination of Scott, the physicians state that "the brain was not large." But this, no doubt, means relatively to the internal capacity of the skull in its then diseased condition. The intermastoid arch, as already noted, shows a remarkably exceptional magnitude of 19 inches, whereas the average of fifty-eight ancient and modern European skulls, as given in the "Thesaurus Craniorum," is only 14·60. The portraits of Wordsworth and Byron show an ample forehead ; and the popular recognition of the "fair large front" of Milton's typical man as the index of superior intellect is an induction universally accepted. But, on the other hand, examples of intellectual greatness undoubtedly occur with the brain little, if at all, in excess of the average size. On the discovery of Dante's remains at Ravenna in 1865, the skull was pronounced to be ample, and exquisite in form. But its actual cubical capacity and estimated brain-weight fall considerably below those of the heaviest ascertained brain-weights of distinguished men. Again, looking at the casts of the skulls of Robert the Bruce and the poet Burns, the first impression is the comparatively small size of head, and the moderate frontal development in each. Mr. Robert Liston, the eminent surgeon, remarked of the former: "The division of the cranium behind the meatus auditorius is large in proportion to that situated before it. The skull is also remarkably wide and capacious in that part, whereas the forehead is rather depressed." * Other characteristics so markedly indicate the elements of physical rather than intellectual vigour, that Mr. Liston expressly pointed out the analogy to "the heads of carnivorous animals." The Bruce was indeed pre-eminently distinguished for courage and deeds of personal prowess; but it was no less by statesmanlike qualities, calm, resolute perseverance, and wise prudence, that he achieved the independence of his country.

Mr. George Combe, the phrenologist, to whom the original cast of Burns' skull was first submitted, thus states the case in reference to the frontal development of the poet: "An unskilful observer looking at the forehead, might suppose it to be moderate in size; but when the dimensions of the anterior lobe, in both length and breadth, are attended to, the intellectual organs will be recognised to have been large. The anterior lobe projects so much that it gives an appearance of narrowness to the forehead which is not real."† The actual dimensions of the skull are, longitudinal diameter, 8 inches; parietal diameter, 5.95; and horizontal circumference, 22·25.

In the year 1865 the bones of Italy's greatest poet, Dante, were submitted to a minute examination under the direction of commissioners appointed by the Italian Government to verify the discovery; and careful measurements were taken of the skull. Dr. H. C. Barlow, describing it from personal observation, says: "The head was finely formed, and the cranium showed, by its ample and exquisite form, that it had held the brain of no ordinary man. It was the most intellectually developed head that I ever remember to have seen. The occipital region was prominently marked, but the frontal was also amply and broadly expanded, and the anterior part of the frontal bone had a vertical direction in relation to the bones of the face." (*Athenæum*, September 9, 1865). But however intellectually developed and exquisite in form the poet's skull may have appeared, the actual measurements fall short of the amplitude here assigned to it. The dimensions were as follows:—Internal capacity, determined by filling the calvarium with grains of rice, 3·1321 lb. avoird., or a

---

* "Archæologia Scotica," Vol. II., p. 450.
† "Phrenological Development of Robert Burns," by George Combe, p. 7.

little over 50 oz. ; circumference, 52 cent. 5 mill.; occipito-frontal diameter, 31 cent. 7 mill.; transverse diameter, taken between the ears, 31 cent. 8 mill.; height, 14 cent. If the internal capacity is accepted without any correction, it would yield 57 oz., but if allowance be made, as in the actual weighing of the brain, for the abstraction of the dura mater and fluids, of say 8 per cent., this would reduce it to about 52·5, or nearly the same weight as that of the mathematician, Gauss. Professor Welcker deducts from 11·6 to 14 per cent., according to the size of the skull; Dr. J. B. Davis recommends a uniform deduction of 10 per cent. If we apply the latter rule, it will reduce the estimated weight of Dante's brain to 51·3 oz.*

Another interesting example of the skull of an Italian poet is that of Ugo Foscolo, a cast of which was taken on the transfer of his remains to the Church of Santa Croce at Florence. Though only fifty years old at the time of his death, the skull was marked by "the entire ossification of the coronal, sagittal, and lambdoidal sutures, and that atrophy of the outer table, manifested by a depression on each side in the posterior half of each parietal, leaving an elevated ridge in the middle, in the position of the sagittal, which is but rarely observed except in extremely advanced age."[†] Sir Henry Holland. who knew the poet intimately, describes him as resembling in temperament the painter Fuseli, "passionately eccentric in social life." Full of genius and original thought, as the writings of Foscolo show him to have been, he "was fiery and impulsive, almost to the verge of madness."[‡] He died in England in obscurity and neglect ; but a regenerated Italy recalled the memory of her lost poet, and transferred his remains to Santa Croce's consecrated soil. The estimated size of his brain is given as 1426 cub. cent., equivalent to 87 cub. in. internal capacity, which corresponds to a weight of brain

* The use of different standards of weights and measures, and of diverse materials for determining the capacity of the skull in different countries, greatly complicates the researches of the craniologist. Some pains have been taken here to bring the various weights and measurements to a common standard. In attempting to do so in reference to the weight of brain of Italy's great poet, the following process was adopted: It was ascertained by experiment that 912·5 grs. of rice, well shaken down, occupied the space of 1000 grs. of water. Hence 3·1321 lbs. rice=3·4324 water. Multiplying this by 1·04, the s.g. of brain, the result is the capacity of the skull, viz., 3·5697 lbs., or 57 oz., as given above. In this and other investigations embodied in the present paper, I have been indebted to the valuable co-operation of my friend and colleague, Prof. H. H. Croft.

† Dr. J. B. Davis, Supp. " Thesaurus Craniorum," p. 7.

‡ Sir H. Holland's " Recollections of Past Life," p. 254.

of 48·44 oz. The longitudinal diameter is 6·90 ; the parietal diameter
5·70 ; the intermastoid arch 15·0 ; and the horizontal circumference
520 m.m., or 20·5 in. The brain capacity of the poet was thus little
more than the European mean deduced by Morton from the miscel-
laneous examples in his collection.

Dr. J. C. Gustav Lucae, in his "Zur Organischen Formenlehre,"
furnishes views and measurements of two other skulls of men of
known intellectual capacity. One of these is Johan Jacob Wilhelm
Heinse, the author of "Ardinghello," a work of high character in the
elements of æsthetic criticism, though as a romance fit to rank with
"Don Juan" in subjective significance and morality. He wrote
another romance entitled "Hildegard ;" in addition to numerous
articles and translations of Petronius, Tasso, &c., which won for him
the high commendation of Goethe, and the more guarded admiration
of Wieland. His skull, as figured by Dr. Lucae, shows the frontal
suture still open at the age of 53, at which he died. The internal
capacity of the skull is stated as 41·4 oz., equivalent to 1173 grms.,
In this, as in other examples hereafter referred to, Dr. Lucae has
gauged the capacity of the skull with peas, and gives the weight in
"unzen." In the results deduced from them here the *unzen* are
assumed to be Prussian ounces, the lb. of 12 oz. equal to 350·78348
grms. Professor Croft has made a series of experiments for me with
a view to correct the error necessarily resulting from the fact that
peas do not entirely fill the cavity. The results show that 82·5
grms. of ordinary sized peas occupy the space of 100 grms. of water.
Deducting 10 per cent. for membranes and fluids, the estimated
brain-weight of Heinse is 1379 grms. or 48·7 oz. av. The dimensions
of the skull are given thus :

	HEIGHT.	LENGTH.	BREADTH.
Fore part......................	4·9	4.0	4·10
Middle part......................	4·10	3·11	5·3
Hind part	3·9	3·6	4·1

The other example produced by Dr. Lucae is that of Dr. Christian
Heinrich Bünger, Professor of Anatomy in the University of

Marburg. In this skull the frontal suture is still more strongly defined at the age of 60 than in that of Heinse. The internal capacity of the skull is stated as 42·8 oz., equivalent to 1213 grms., which, dealt with as above stated, yields 1410 grms. or 49·8 oz. av. Other dimensions of the skull are given as follows :

	HEIGHT.	LENGTH.	BREADTH.
Fore part......................	4·8	4·1	4·20
Middle part	4·9	4·1	5·0
Hind part......................	3·7	3.10	4·1

Professor Welcker assigns a standard, which was accepted by Dr. Thurnam, thus : "Skulls of more than 540 to 550 millemetres in horizontal circumference (the weight of brain belonging to which is 1490 to 1560 grms., or 52·5–55 oz. avoirdupois), are to be regarded as exceptionally large. The designation of *kephalones*, proposed by Virchow, might commence from this point. Men with great mental endowments fall, for the most part, under the definition of kephalony. If we consider the relations of capacity, 1800 grms. (63·5 oz.) appears to be the greatest attainable weight of brain within a skull not pathologically enlarged." But the brain of Cuvier—the heaviest healthy brain yet recorded,—exceeded this. Its weight is stated by Wagner as 1861 grammes, or 65·8 oz. ; but this M. Broca corrects to 1829·96 grammes. Even thus reduced it exceeds the limits assigned by Professor Welcker to the normal healthy brain. But a curious commentary upon this is furnished by the fact that the modern English skull which Dr. J. B. Davis selects as presenting the most striking analogy to the Neanderthal skull—"the most ape-like skull which Professor Huxley had ever beheld,"—though marked not only by the prominence of the superciliary ridges, but by great depression of the frontal region, appears to have a cubical capacity equivalent to that of Dr. Abercrombie, whose brain is only surpassed by that of Cuvier among the ascertained brain-weights of distinguished men.* Its capacity is 94 oz. of sand, or 113 cubic inches,

---

equivalent—after making the requisite deduction for membranes
and fluids,—to a brain-weight of 63 oz.

I have attempted in the following table to reduce to some common
standard such imperfect glimpses as are recoverable of the cranial
capacity of some distinguished men, of whose actual brain-weights no
record exists :

TABLE III.

CRANIAL CAPACITY OF DISTINGUISHED MEN.

	LENGTH.	BREADTH.	CIRCUMFERENCE	ESTIMATED AIN-WEIGHT.
Dante	....	....	....	51·3
Robert the Bruce....	7·70	6·25	22·25	....
Burns	8·00	5·95	22·25	....
Scott (head)........	9·	6·40	23·10	....
Heinse...............	....	5·30	....	48·7
Bünger	....	5·00	....	49·8
Ugo Foscolo........	6·90	5·70	20·50	48·4

Some of the examples adduced in the above table appear to
exhibit instances of mental endowment of high character, without the
corresponding degree of cranial, and consequently cerebral development. The following table exhibits recorded examples of a series
of actual brain-weights of distinguished men. It seems to lend confirmation to the idea that great manifestation of mental endowment
is correlated, in the majority of observed cases, to a brain above the
normal average in mass or weight. But even here intellect and
brain-weight are not strictly in uniform ratio. Several of the following brain-weights, including that of Tiedemann, are furnished by
Wagner, in the "Vorstudien des Menschlichen Gehirns;" but in
an elaborate table of brain-weights given in the "Morphologie und
physiologie des Menschlichen gehirns als Seelenorgan," the brain of
Byron is classed above all except Cuvier; while Vogt gives the same
place, by estimate, to Schiller's, as next in rank to that of the great
naturalist among highly developed brains. Dr. Thurnam states his
authorities for others, when producing them in his valuable contribution to the *Journal of Mental Science* "On the Weight of the
Brain." For that of Webster he refers to "the unsatisfactory
article on the brain of Daniel Webster, *Edin. Med. Surg. Journ.*,
vol. lxxix., p. 355." Dr. J. C. Nott, in his "Comparative Anatomy

of Races" ("Types of Mankind," p. 453), says: "Dr. Wyman, in his post-mortem examination of the famed Daniel Webster, found the internal capacity of the cranium to be 122 cubic inches, and in a private letter to me, he says: 'The circumference was measured outside of the integuments before the scalp was removed, and may, perhaps, as there was much emaciation, be a little less than in health.' It was 23¾ inches in circumference; and the Doctor states that it is well-known there are several heads in Boston larger than Webster's. I have myself, in the last few weeks, measured half a dozen heads as large and larger." The circumference, it will be seen, exceeds the corresponding measurement of Scott's head, taken under similar circumstances. But the statement of 122 cubic inches as the internal capacity of Webster's skull seems open to question. If correct, instead of 53·5 oz. of brain-weight, as stated in the following table, it is the equivalent of a brain-weight of fully 65 oz., or one in excess even of that of Cuvier. The brain-weights of Goodsir, Simpson and Agassiz, are given in the following table from the reported autopsy in each case.

TABLE IV.
BRAIN-WEIGHTS OF DISTINGUISHED MEN.

			AGE.	Oz.	GRMS.
1	Cuvier...............	Naturalist	63	64·5	1830
2	Byron	Poet	36	63·5 ?	1799
3	Abercrombie........	Philosopher, Physician..	64	63·	1785
4	Schiller.	Poet	46	63· ?	1785
5	Goodsir...........	Anatomist	53	57·55	1629
6	Spurzheim..........	Phrenologist, Physician..	56	55·06	1559
7	Simpson............	Physician, Archæologist..	59	54·	1530
8	Dirichlet...........	Mathematician	54	53·6	1520
9	De Morny..........	Statesman	50	53·6	1520
10	Daniel Webster.....	Statesman	70	53·5	1516
11	Campbell	Lord Chancellor........	80	53·5	1516
12	Agassiz............	Naturalist.............	66	53·4	1512
13	Chalmers	Author, Preacher.......	67	53·	1502
14	Fuchs	Pathologist......... ...	52	52·9	1499
15	Gauss	Mathematician	78	52·6	1492
16	Dupuytren.........	Surgeon	58	50·7	1436
17	Whewell	Philosopher	71	49·	1390
18	Hermann	Philologist	51	47·9	1358
19	Tiedemann	Physiologist...........	80	44·2	1254
20	Hausmann..........	Mineralogist...........	77	43·2	1226

3

Dr. Thurnam, in producing fifteen of the above examples, remarks : "Altogether, they decidedly confirm the generally received view of the connection between size of brain and mental power and intelligence :" and he adds his conviction that if the examination of the brain in the upper ranks of society, and in men whose mental endowments are well known, were more generally available, further confirmation would be given to this conclusion. The converse, at least, is certain, that no great intelligence or unwonted mental power is possible with a brain much below the average in mass and weight. But there are unquestionable indications that a large, healthy brain may exist without the manifestation of great mental power ; while brains inferior both in size and weight have been the organs of unwonted intelligence and mental activity.

In the "Philosophical Transactions" of 1861, Dr. Boyd published an elaborate series of researches illustrative of the weight of various organs of the human body, including the weights of 2,000 brains. Most of the healthy brains are those of patients in the St. Marylebone Infirmary, and have already been referred to as necessarily representing the indigent and uneducated classes of London. Here, therefore, if an unusually large brain is the index of intellectual power, every probability was against the occurrence of brains above the average size or weight. But the results by no means confirm this assumption. Among the patients in the Edinburgh Royal Infirmary, in like manner, though including the better class of artizans and others from country districts, we might still look for a confirmation of M. Broca's assumption, based on extensive observations of French crania, "that, other things being equal, whether as the result of education, or by hereditary transmission, the volume of the skull, and consequently of the brain, is greater in the higher than in the lower classes." But Dr. Peacock's tables include four brain-weights, three of them of a sailor, a printer, and a tailor, respectively, ranging from 61 to 62·75 oz. ; and so surpassing all but two, or at the most three, of the heaviest ascertained brain-weights of distinguished men. Tried by the posthumous test of internal capacity, three skulls of nameless Frenchmen, derived from the common cemeteries of Paris, in like manner showed brains equalling in size that of Cuvier. The following are the maximum brain-weights among the St. Marylebone patients apparently unaffected by cerebral disease.

TABLE V.

MAXIMUM BRAIN-WEIGHTS—ST. MARYLEBONE.

AGE.	MALE.		FEMALE.	
	Oz.	Grms.	Oz.	Grms.
7—14	57·25	1622	52·	1473
14—20	58·5	1658	52·	1473
20—30	57·	1615	55·25	1565
30—40	60·75	1721	53·	1502
40—50	60·	1700	52·5	1488
50—60	59·	1672	52·5	1488
60—70	59·5	1686	54·	1530
70—80	55·25	1565	49·5	1403
80—	53·75	1523	48·	1360
All Ages. 7—80	60·75	1721	55·25	1565

The stature, or relative size of body, has already been referred to as an element in testing the comparative male and female weight of brain; and it is one which ought not to be overlooked in estimating the comparative size and weight of the brains of distinguished men. From my own recollections of Dr. Chalmers, who was of moderate stature, his head appeared proportionally large. The same was noticeable in the cases of Lord Jeffrey, Lord Macaulay, Sir James Y. Simpson, and very markedly so in that of De Quincey. The philosopher Kant was also of small stature; and Dr. Thurnam refers to the observation of Carus that he had a head not absolutely large, though, in proportion to the small and puny body of that eminent thinker, it was of remarkable size. Among the large-brained artizans of the Marylebone Infirmary, on the contrary, the probabilities are in favour of a majority of them being men of full muscular development and ample stature. Nevertheless, with every allowance for this, it still remains probable, if not demonstrable, that from the same humble and unnoted class, examples of megalocephaly could be selected little short in cerebral mass, and apparently in brain-weight, of the group of men whose large brains are recognized as the concomitants of exceptionally great mental capacity and intellectual vigour. Unless, therefore, we are contented to accept the poet's dictum, " Their lot forbad,"* and assume that "chill penury repressed their noble rage, and

* Gray's " Elegy."

froze the genial current of the soul," it is manifest that other elements besides those of volume or weight are essential as cerebral indices of mental power. Dr. Thurnam, after noting examples that had come under his own notice of brain-weights above the medium—but which, as those of insane patients, may be assigned to other causes than healthy cerebral development,—adds: "The heaviest brain weighed by me (62 oz., or 1760 grms.) was that of an uneducated butcher, who was just able to read, and who died suddenly of epilepsy, combined with mania, after about a year's illness. The head was large, but well-formed; the brain of normal consistence; the *puncta vasculosa* numerous." In cases like this, of weighty brain with no corresponding manifestation of intellectual power, something else was wanting besides a less circumscribed sphere. The mere position of a humble artizan or labourer will not suffice to mar the capacity to "make by force his merit known," which pertains to the "divinely gifted man."

Arkwright, Franklin, Watt, Stephenson, and others of the like type of self-made men, are not rare. Among those large-brained artizans, scarcely one can have had a more limited sphere for the exercise of mental vigour than the poet Burns, the child of poverty and toil, who refers to his own early years as passed in "the unceasing moil of a galley-slave." In his case the very means essential to a healthy physical development were stinted at the most critical period of life. His brother Gilbert says: "We lived sparingly. For several years butcher's meat was a stranger to the house; while all exerted themselves to the utmost of their strength, and rather beyond it, in the labours of the farm. My brother, at the age of thirteen, assisted in thrashing the crop of corn, and at fifteen was the principal labourer on the farm." Such premature toil and privations left their permanent stamp on his frame. "Externally, the consequences appeared in a stoop of the shoulders, which never left him; but internally, in the more serious form of mental depression, attended by a nervous disorder which affected the movements of the heart." He had only exchanged the toil on his father's farm for equally unremitting labour on his own, when the finest of his poems were written; nor would it be inconsistent with all the facts to assume that the privations of his early life diminished his capacity for continuous mental activity; as it undoubtedly impaired his physical constitution. But, while the possession of a brain much above the average in size might have seemed to account for his triumph over the depressing influences of his limited sphere, the fact that his brain appears to have been rather below than above the average size, points to some other requisite than mere cerebral mass as essential to intellectual vigour.

The brain is influenced in all its functions by the character and the amount of blood circulating through it, and promptly manifests the effects of any deleterious substance, such as alcohol or opium, introduced into its tissues. It depends, like other portions of the nervous system, on an adequate supply of nourishment. In both respects the brain of the Ayrshire poet was injuriously affected, in so far as we may infer from all the known circumstances of his life.

The human brain is large in proportion to the body in infancy and youth; and the opinions of leading anatomists and physiologists early in the present century favoured the idea that it attained its full size within a few years after birth. Professor Soemmering assumed this to take place so early as the third year. Sir William Hamilton explicitly stated his conclusion thus : " In man the encephalon reaches its full size about seven years of age';" and Tiedemann assigns the eighth year as that in which it attains its greatest development. But the more accurate and extended observations since carried on rather tend to the conclusion that the brain not only goes on increasing in size and weight to a much later period of life, but that, under exceptionally favouring circumstances, it may increase in weight long after the body has attained its maximum.

The largest average brain-weights, as determined by observations on the brains of upwards of two thousand men and women in different countries of Europe, have indeed been found in those not above twenty years of age ; and from a nearly equal number of English examples, Dr. Boyd determines the period of greatest average weight to be the interval between fourteen and twenty years of age ; but this includes cases in which death has ensued from undue or premature brain development.

Other evidence leaves no room for doubt that cases are not rare of the growth, or increased density of the brain up to middle age ; while the observations of Professor Welcker indicate this process extended to a later period of life. The average brain-weights, as given by Boyd, Peacock, and Broca, from healthy or sane cases, along with those of Welcker, include the weights of forty-seven male brains from ten to twenty years of age, giving an average of

49·6 oz., or 1405 grms.; and of one hundred and twelve male brains from twenty to thirty years of age, giving an average of 48·9 oz., or 1384 grms.; and the results of a nearly equal number of female brains closely approximate. They embrace English, Scotch, German, and French, men and women. Dr. Welcker's results indicate the period of maximum brain-weight to be between 30-40, as shewn in the following table :

TABLE VI.

AVERAGE WEIGHT OF THE BRAIN AT DIFFERENT AGES.

	MALE.		FEMALE.	
	Oz. Av.	Grms.	Oz. Av.	Grms.
From 10—20	47·5	1346	43·1	1221
20— 30	49·5	1404	44·1	1251
30—40	49·5	1404	44·8	1272
40—50	48·6	1379	43·5	1234
50—60	48·1	1365	43·5	1234
60—70	46·1	1306	42·8	1213

In the female examples, amounting to thirty-one between seventy and eighty years of age, and six between eighty and ninety, the continuous diminution of brain-weight corresponds with the increasing age ; but in the male examples, sixty-five cases between sixty and seventy years of age yield an average brain-weight of 46·1 oz., while twenty-seven cases between seventy and eighty years of age give 47·9 as the average ; falling in the next decade to 43·8.

It may be inferred from the number of cases pointing to an early attainment of the highest average brain-weight, not that the brain differs from all other internal organs of the human body in attaining its maximum before the period of puberty, but that physical as. well as mental vigour are dependent on the maintenance of a nice equilibrium between the brain and the other organs while in process of development. The observations of Dr. Boyd, including the results of 2,614 *post mortem* examinations of sane and insane patients of all ages, showed that the average weight of the brain of " still-born " children at the full period was much greater than that of the newborn living child. It is a legitimate inference, therefore, that death in the former cases was traceable to an excessive premature development of the brain. Again, when it is shown from numerous cases that the highest average weights of brain in both sexes occur not later than twenty years of age, it appears a more legitimate inference to trace to exceptional cerebral development towards the period of adolescence, the mortality which rendered available so many examples of unusually large or heavy brains, than to assume that the normal healthy brain begins to diminish at that age.

It is a fact familiar to popular observation that a large head in youth is apt to be unfavourable to life. A tendency to epilepsy appears to be the frequent concomitant of an unusually large brain ; and with the congestion accompanying its abnormal condition, this may account for the weights of such diseased brains as have been repeatedly found in excess of nearly all the recorded examples of megalocephaly in the cases of distinguished men. But a greater interest attaches to a remarkable example of healthy megalocephaly recorded in the *British Medical Journal* for 1872. The case was that of a boy thirteen years of age, who died in Middlesex Hospital from injuries caused by a fall from an omnibus. His brain was found to weigh 58 oz. He had been a particularly healthy lad, without any evidence of rachitis, and very intelligent. This is a strikingly exceptional case of a healthy brain, at the age of thirteen, exceeding in weight all but two of the greatest ascertained brain-weights of distinguished men.

From the evidence already adduced of relative cubical capacity of the skulls of different races, it appears, as was to be expected, that there is a greater prevalence of the amply-developed brain among the higher and more civilized races. But all averages are apt to be deceptive; and the progressive scale from the smallest up to the greatest mass of brain is by no means in the precise ratio of an intellectual scale of progression. The results of Dr. J. B. Davis's investigations, based on the study of a large, and in many cases a seemingly adequate number of skulls, bring out this remarkable fact, that, so far from the Polynesians occupying a rank in the lowest scale, as affirmed by Professor Vogt, the Oceanic races of the Pacific generally rank in internal capacity of skull, and consequent size of brain, next to the European.

But it is of more importance for our present enquiry to note that, as exceptionally large and heavy brains occur among the most civilized races, in some cases—and in some only,—accompanied with corresponding manifestations of unusual intellectual power : so also it

becomes apparent that skulls much exceeding the average, and some
of remarkable internal capacity, are met with among barbarian races,
and even among some of the lowest savages. Taking the crania
in the elaborate series of tables in Dr. J. B. Davis's "Thesaurus
Craniorum," with an internal capacity above 100 cubic inches, they
will rank in order as follows :

Chinese	111·8
Maduran	110.6
Marquesan	110.6
Kanaka	108·8
Javan	107·
Negro	105·8
Australian	104·5
Kafir	104·5
Bakele	103·3
Tidorese	103·3
Bhotia	102·7
Bodo	100·9
Hindoo	100·9
Sumatran	100·9

Among the European series the largest is an Irish cranium of
121·6 cubic in., and next to it comes an Italian, 114·3, and an
Englishman, 112·4; an ancient Briton from a Yorkshire Long Barrow, 109·4 : an ancient Roman, 106·4 ; a Lapp, 105·8 ; an ancient
Gaul, 103·7 ; a Briton of Roman times, 103·3 ; a Merovingian
Frank, 101·5 ; and an Anglo-Saxon, 100·9. Those and other examples of the like kind are full of interest as showing the recurrence
of megalocephalic variations from the common cranial and cerebral
standard among ancient races ; and among rudest savages as well
as among the most cultivated classes of modern civilized nations.
But the order shown in the above instances is derived from purely
exceptional examples, and is no key to the relative capacity of the
races named.

Opportunities for testing the size and weight of the brain among
barbarous races are only rarely accessible to those who are qualified
to avail themselves of them for the purposes of science. Some near
approximation to the relative brain-weight of the English, Scotch,
German, and French, may now be assumed to have been established.
Dr. Thurnam instituted a comparison between those and two of the
prehistoric races of Britain—the Dolichocephali of the Long Barrows,

and the Brachycephali of the Round Barrows of England.* The
results are curious, as showing not only a greater capacity in the
ancient British skulls than the average modern German, French, or
English head; but an actual average higher than that of all but five
of the most distinguished men of Europe, whose brain-weights have
been recorded. On comparing the ancient skulls with those of
modern Europeans, as determined by gauging the capacity of both
by the same process, the following are the results presented, according
to the authorities named :

TABLE VII.

SKULLS OF MEN.	No.	Weight of Sand.	Cubic Inches.	Capacity. Centimetres.	Brain-weight oz. av.
Ancient Britons, L. Barrows ..	18	82	99	1622	54.
" " R. Barrows ..	18	80½	98	1605	53.5
Modern English, *Morton*	28	77	94	1540	52.2
" French, *Broca*	357	74	91	1502	50.6
" German, *Welcker*	30	72	88	1450	49.

The highest average of any nationality, as determined by Drs.
Reid and Peacock from the weighing of 157 brains of male patients,
chiefly Scottish Lowlanders, in the Royal Infirmary of Edinburgh,
is little more than 50 oz., or 1417 grammes ; whereas the estimated
average brain-weight in the ancient British skulls is 54 oz. for the
Dolichocephali of the Long Barrows, which equals that of Sir James
Simpson, and exceeds all but six of the most distinguished men.
For the Brachycephali of the Round Barrows it is 53·5 oz., which is
in excess of the brain-weights of Agassiz, Chalmers, Whewell, and
other distinguished men, and exactly accords with that of Daniel
Webster and Lord Chancellor Campbell. In so far, moreover, as
this illustrates the cerebral capacity of ancient races, it is in each
case an average obtained by gauging eighteen skulls, and not the
cranial capacity of one or two exceptionally large ones. Dr. Thurnam does indeed suggest that the Barrows may have been the sepulchres of chiefs; nor is this unlikely; but the superior vigour and
mental endowment which this implies fails to account for a cerebral
capacity surpassing all but the most distinguished men of science and
letters in modern Europe. Rather may we conclude from this, as

from other evidence, that quality of brain may, within certain limits, be of more significance than mere quantity; and that brains of the same volume, and agreeing in weight, may greatly differ in minute structure and in powers of cerebration.

In the case of the ancient British Barrow Builders we seem to have large heads and remarkable development of brain, without any indications of an equivalent in intellectual power; and although the estimated brain-weight derived from gauging the capacity of the empty chamber of the skull proceeds on the assumption of mass and weight agreeing, sufficient data exist to justify the adoption of this for approximate results. The average weight of brain of twelve male Negroes of undetermined tribes, deduced from gauging their skulls, has been determined at 1255 grammes, or 44·3 oz. The actual weight of brain of the Negro of Guinea described by Professor Calori, was 1260 grammes. Other examples vary considerably from the average. Mascagni gives 1458 grammes as the weight of one Negro brain weighed by him; equivalent to an actual brain-weight of 51·5 oz., which is greater than that of Dupuytren, Whewell, Hermann, or Tiedemann. Nevertheless, although the extremes are great, and are confirmed by a like diversity in measurements of the horizontal circumference and of internal capacity, the average result given above appears to be a fair and reliable one. But the same process, when applied to determine the comparative cranial capacities of the native American races, discloses results of a wholly different character, and widely at variance with those above described relating to the ancient races of Britain. On the continent of America the native ethnical scale embraces a comparatively narrow range; and any intrusive elements are sufficiently recent to be easily eliminated. The Patagonian and the Fuegian rank alongside of the Bushman, the Andaman Islander, or the Australian, as among the lowest types of humanity; while the Aztecs, Mayas, Quichuas, and Aymaras, attained to the highest scale which has been reached independently by any native American race. We owe to the zealous and indefatigable labours of Dr. Morton, alike in the formation of his great collection of human crania, and in the published results embodied in the "Crania Americana," the chief knowledge derived from this class of evidence in reference to the races of the New World. In one respect, at least, those results stand out in striking contrast to the large-headed barbarian Barrow Builders of ancient Britain. Dr.

Morton subdivides the American races into the Toltecan Race, embracing the semi-civilized communities of Mexico, Bogota, and Peru, and the barbarous tribes scattered over the continent from the Arctic Circle to Tierra del Fuego. His latest views are embodied in a contribution to Schoolcraft's "History of the Indian Tribes of the United States," entitled "The Physical Type of the American Indians." In treating of the volume of brain, he draws special attention to the Peruvian skulls, 201 in number, obtained for him from the cemeteries of Pisco, Pachacamac, and Arica. "Herera informs us that Pachacamac was sacred to priests, nobles, and other persons of distinction; and there is ample evidence that Arica and Pisco, though free to all classes, were among the most favoured cemeteries of Peru." Dr. Morton accordingly adds: "It is of some importance to the present inquiry, that nearly one-half of this series of Peruvian crania was obtained at Pachacamac; whence the inference that they belonged to the most intellectual and cultivated portion of the Peruvian nation; for in Peru learning of every kind was an exclusive privilege of the ruling caste." In reality, however, the latest additions to our knowledge of the physical characteristics of the ancient Peruvians tend to confirm the idea of the existence of two distinct races : a patrician order occupying a position analogous to the Franks of Gaul or the Normans of England, though more aptly to be compared to the Brahmins of India; and a more numerous class, constituting the labouring and industrial orders of the community, abundantly represented in the Pacific Coast tribes of Peru, the cemeteries of which have furnished the larger number of crania to European and American collections.

To such a patrician order or cast the intellectual superiority and privileges of the governing race pertained. But whatever may have been the exclusive prerogatives of the patrician and sacerdotal orders, there is no doubt that the Peruvians as a people had carried metallurgy to as high a development as has been attained by any race ignorant of working in iron. They had acquired great skill in the arts of the goldsmith, the engraver, chaser, and modeller. Pottery was fashioned into many artistic and fanciful forms, showing ingenuity and great versatility of fancy. They excelled as engineers, architects, sculptors, weavers, and agriculturists. Their public works display great skill, combined with comprehensive aims of practical utility; and alone, among all the nations of the New World, they had domesticated animals, and trained them as beasts of burden. It is not, therefore, without reason that Dr. Morton adds : " When we consider the institutions of the old Peruvians, their comparatively advanced civilization, their tombs and temples, mountain roads and monolithic gateways, together with their knowledge of certain ornamental arts, it is surprising to find that they possessed a brain no larger than the Hottentot and New Hollander, and far below that of the barbarous hordes of their own race. For, on measuring 155 crania, nearly all derived from the sepulchres just mentioned, they give but 75 cubic inches [equivalent, after due deduction for membranes and fluids, to a brain of 40·1 oz. av. in weight,] for the average bulk of the brain. Of the whole number, only one attains the capacity of 101 cubic inches, and the minimum sinks to 58, the smallest in the whole series of 641 measured crania. It is important further to remark that the sexes are nearly equally represented, viz., eighty men and seventy-five women."

Other collections subsequently formed have largely added to our means of testing the curious question thus raised of the apparent inverse ratio of volume of brain to intellectual power and progressive civilization among the native races of the American continent. In 1866, Mr. E. G. Squier presented to the Peabody Museum of American Archæology and Ethnology at Harvard, a collection of seventy-five Peruvian skulls, obtained by himself from various localities both on the coast and in the interior. "The skulls from the interior represent the Aymara on Lake Titicaca, as well as the Quichua, Cuzco, or Inca families; and the skulls of every coast family from Tumbes to Atacama, or from Ecuador to Chili."* Subsequently the curator, the late Professor Jeffreys Wyman, made this collection, along with two others, of skulls from the mounds of Kentucky and Florida, the subject of careful comparative measurements. The following are the results : The crania from Florida were chiefly obtained from a burial place near an ancient Indian shell mound of gigantic proportions, a few miles distant from Cedar Keys. They are eighteen in number, and have a mean capacity of 1375·7 cubic centimetres, or nearly 84 cubic inches. The skulls from the Kentucky mounds, twenty-four in number, show a mean capacity of 1313 cubic centimetres, 80·21 cubic inches, with a difference of 125 cubic centimetres, or 7·61 cubic inches in favour of the males. Yet, small as the Kentucky skulls

---

* "Peabody Museum Annual Report, 1868," p. 7.

are, they exceed the Peruvian ones. Keeping in view the varied sources of the latter, Professor Wyman remarks : "Although the crania from the several localities show some differences as regards capacity, yet in most other respects they are alike." And the numbers, when viewed separately, are too few to attach much importance to variations within so narrow a range. Nevertheless it is noteworthy that the highest mean is that of the Aymaras of Lake Titicaca; and this difference is considerably increased by measurements derived from subsequent additions to the Harvard collection, received since the death of Professor Wyman from the high valley of Lake Titicaca. In other respects besides their marked superiority in size, the latter crania differ from those of the Coast tribes, and confirm the earlier deduction of an ethnical distinction between the more numerous race so abundantly represented in the coast cemeteries, and that which is chiefly represented by crania brought from the interior. The numbers from the several localities selected by Professor Wyman as fair average specimens of the whole stand thus: six from burial towers, or chulpas, near Lake Titicaca, 1292; five from Cajamaquilla, 1268·75; fourteen from Casma, 1254; four from Truxillo, 1236; four from Pachicamac, 1195; sixteen from Amacavilca, 1176·2; and seven from Grand Chimu, 1094·28.

In 1872, the collection of Peruvian crania in the Peabody Museum was augmented by a large addition from 330 skulls obtained by Professor Agassiz, through the intervention of Mr. T. J. Hutchinson, British Consul at Callao, in Peru. From those contributed to the Harvard Museum, Dr. Wyman selected eleven as apparently the only ones unaffected by any artificial compression or distortion, and therefore valuable as illustrations of the normal shape of the Peruvian head. They are quite symmetrical. The occiput, instead of being flattened or vertical, as in the distorted crania, has the ordinary curves, and in some of them is prominent. Two of them are marked by a low, retreating forehead; but in all the others the forehead is moderately developed. As, moreover, the larger half appear to be the skulls of females, this accounts for the mean capacity falling below the Peruvian average. But they are all small. The largest of them is only 1260 cub. cent., or less than 74 cub. in.; and the average capacity of ten of them is 1129 cub. cent., or 69 cub. in.

The collection, as a whole, differs from that of Mr. Squier, in having been derived from the huacas, or ancient graves of one locality, that of Ancon, near Callao. Professor Wyman stated as the result of his careful study of them: "The average capacity obtained from the whole collection, including those having the distorted as well as the natural shape, varies but little from that of previous measurements," including those of Morton and Meigs, and his own results from the Squier collection.

Another collection of one hundred and fifty ancient skulls, obtained by Mr. Hutchinson during his residence in Peru, and presented to the Anthropological Institute of London, has the additional value, like that of Mr. Squier, of having been carefully selected from different localities, including Santos, Ica, Ancon, Passamayo, and Cerro del Oro; and the same may be said of those enumerated in the "Thesaurus Craniorum" of Dr. J. B. Davis. We have thus unusually ample materials for determining the cranial characteristics of this remarkable people, and the results in every case are the same. After a careful examination of the Peruvian skulls, in the London anthropological collection, Professor Busk states his conclusions thus: "The mean capacity of the larger skulls, which may be regarded as males, appears, as far as I have gone, to be about 80 cubic inches, equivalent to a brain of about 45 ounces, roughly estimated. This capacity, and the measurements above cited, show that the crania generally are of small size;" and he adds: "this is in accord with the statements of all observers." *

Dr. J. B. Davis has added to the valuable data included in his "Thesaurus Craniorum," a series of measurements of skeletons. Unfortunately that of a male Quichua, procured by him in the form of a "Peruvian mummy," proved to be affected with carious disease about the last dorsal and upper lumbar vertebræ; and consequently the length of the vertebral column essential for comparison with the skeletons of other races, is wanting; but the other measurements indicate in this example a stature below the average, while the skull exceeds it. The average internal capacity of eighteen Quichua male skulls, as given by Dr. J. B. Davis, is 73, whereas this is 78·5. That the ancient Peruvian skulls are, with rare exceptions, of small size, is undoubted; and in view of this it becomes a matter of some importance to determine whether this was in any degree due to a correspondingly small stature. Obscure references

---

* "Journal of Anthropol. Inst.," Vol. III., p. 92.

are found in the legendary history of Peru to a pigmy race. Pedro de Cieza de Leon, whose travels have been translated by Mr. Markham, refers to the first emigration of the Indians of Chincha to that valley, "where they found many inhabitants, but all of such small stature, that the tallest was barely two cubits high" (p. 260). Garcilasso de la Vega repeats another tradition heard by himself in Peru, of a race of giants who came by sea to the country, and were so tall that the natives reached no higher than their knees. They lived by rapine, and wasted the whole country till they were destroyed by fire from heaven. Traditions of this class may possibly point to the existence of an aboriginal race of small stature. The aborigines of Guatemala, Salvador, and Nicaragua, are described as below the middle size (Bancroft, Vol. I., p. 688); and Von Tchudi divides the wild Indians of Peru into the Iscuchanos, the natives of the highlands, a tall, slim, vigorous race, with the head proportionally large and the forehead low; and those of the hot lowlands, a smaller race, lank, but broad shouldered, with a broad face and small round chin. There appear, therefore, to be traces of one or more aboriginal races of small stature. But Dr. Morton says expressly of the Peruvians : " Our knowledge of their physical appearance is derived solely from their tombs. In stature they appear not to have been in any respect remarkable, nor to have differed from the cognate nations except in the conformation of the head, which is small, greatly elongated, narrow its whole length, with a very retreating forehead, and possessing more symmetry than is usual in skulls of the American race." Some of the characteristics here referred to are, in part at least, the result of artificial modifications; but the small head appears to be an indisputable characteristic of the most numerous ancient people of Peru.

It may not unreasonably excite surprise that Dr. Morton should have adduced results apparently pointing to the conclusion that civilization had progressed among the native races of the American continent in an inverse ratio to the volume of brain; and yet passed it over with such slight comment. The only hint at a solution of the difficulty is where, as he draws his work to a close, he indicates the recognition of a greater anterior and coronal development in the smaller Peruvian brain. "It is curious," he says, "to observe that the barbarous nations possess a larger brain by five and a half cubic inches, than the Toltecans; while, on the other hand, the Toltecans

possess a greater relative capacity of the anterior chamber of the
skull in the proportion of 42·3 to 41·8. Again, the coronal region,
though absolutely greater in the barbarous tribes, is rather larger in
proportion in the demi-civilized tribes."* But Dr. Morton also noted
that the heads of nine Peruvian children in his possession "appear to
be nearly if not quite as large as those of children of other nations
at the same age;"† so that he seemed to recognize something equivalent to an arrested cerebral development accompanying the intellectual
activity of this remarkable people at some later stage, yet without
apparently affecting their mental power. But it was characteristic
of this minute and painstaking observer to accumulate and set forth
his results, unaffected by any apparent difficulties or inconsistencies
which they might seem to involve. In summing up his investigations "On the internal capacity of the cranium in the different races
of men," he thus concludes:‡ "Respecting the American race, I
have nothing to add, excepting the striking fact that of all the
American nations, the Peruvians had the smallest heads, while those
of the Mexicans were something larger, and those of the barbarous
tribes the largest of all," viz.:

Toltecan Nations { Peruvians, collectively... 75 cub. inches.
　　　　　　　　　{ Mexicans,　　"　　... 79　"　　"
Barbarous Tribes........................ 82　"　　"

The enlarged tables given in the catalogue of Dr. J. Aitken Meigs,
increase this inverse ratio of cerebral capacity, thus:

Peruvians .... 75·3
Mexicans .... 81·7
Barbarous Tribes .... 84·0

"The great American group," he says, "is, in several respects, well
represented in the collection. It includes 490 crania and 13 casts,
making a total of 503 from nearly 70 different nations and tribes.
Of this large number 256 belong to the Toltecan race [embracing the
semi-civilized communities of Mexico, Bogota and Peru,] and 247 to
the barbarous tribes scattered over the continent. Of 164 measurements of crania of the barbarous tribes, the largest is 104 cubic
inches; the smallest 69; and the mean of all 84. One hundred and
fifty-two Peruvian skulls give 101 cubic inches for the largest internal
capacity, 58 for the smallest, and 75·3 for the average of all."§

* "Crania Americana," p. 260.
† "Crania Americana," p. 132.
‡ "Crania Americana," p. 261.
§ "Introductory Note, Catalogue," p. 10.

The results which Professor Jeffreys Wyman arrived at from a careful comparative measurement of the Squier collection, were confirmed by his subsequent study of that of Professor Agassiz, and may be quoted as applying to both; for he sums up his later investigations with the remark: "These results agree with all previous conclusions with regard to the diminutive size of the ancient Peruvian brain."* Of the Squier collection he says: "The average capacity of the fifty-six crania measured agrees very closely with that indicated by Morton and Meigs, viz., 1230 centimetres, or 75 cub. inches, which is considerably less than that of the barbarous tribes of America, and almost exactly that of the Australians and Hottentots as given by Morton and Meigs, and smaller than that derived from a larger number of measurements by Davis. Thus we have, in this particular, a race which has established a complex civil and religious polity, and made great progress in the useful and fine arts—as its pottery, textile fabrics, wrought metals, highways and aqueducts, colossal architectural structures and court of almost imperial splendour prove,—on the same level, as regards the quantity of brain, with a race whose social and religious conditions are among the most degraded exhibited by the human race. All this goes to show, and cannot be too much insisted upon, that the relative capacity of the skull is to be considered merely as an anatomical and not as a physiological characteristic; and unless the quality of the brain can be represented at the same time as the quantity, brain measurement cannot be assumed as an indication of the intellectual position of races any more than of individuals."†

The only definite attempt which Dr. Morton made to solve the difficulty thus presented to us, curiously evades its true point. "Something," he says, "may be attributed to a primitive difference of stock; but more, perhaps, to the contrasted activity of the two races." Here, however, it is not a case of intellectual activity accompanied by, and seemingly begetting an increased volume of brain; but only the assumption of greater activity in the small-brained race to account for its triumph over larger-brained barbarous tribes in the attainment of numerous elements of a native-born civilisation. The question is, how to account for this intellectual activity, with all its marvellous results, attained by a race with an

---

* "Peabody Museum Report, 1874," p. 10.
† "Peabody Museum Report, 1871," p. 11.

average brain of no greater volume than that of the Bushman, the Australian, or other lowest types of humanity.

The Nilotic Egyptian race, of composite ethnical character, presents striking elements of comparison, in the ingenious arts and constructive skill of the ancient dwellers in the Nile valley; but whether we take the Egyptian of the Catacombs, the Copt, or the Fellah, we seek in vain for like microcephalous characteristics. Among modern races the Chinese exhibit many analogies in arts and social life to the ancient Peruvians. But their cerebral capacity presents no correspondence to that of the American race. Dr. Morton gives a mean capacity for the Chinese skull of 85, as compared with the Peruvian 75·3, while Dr. Davis derives from nineteen skulls a mean internal capacity of 76·7 oz. av., or 93 cubic in.

But another Asiatic race, that of the Hindoos—also associated with a remarkable ancient civilization, and a social and religious organization not without suggestive analogies both to ancient Egypt and Peru,—is noticeable for like microcephalous characteristics. In completing the anatomical measurements with which Dr. Morton closes his great work, he places the Ethiopian lowest in the scale of internal capacity of cranium; but, while including the Hindoo in his Caucasian group, he adds: "It is proper to mention that but three Hindoos are admitted in the whole number, because the skulls of these people are probably smaller than those of any other existing nation. For example, seventeen Hindoo heads give a mean of but 75 cubic inches."* The Vedahs of Ceylon, the Mincopies, the Negritos, and the Bushmen, appear to vie with the Hindoos in smallness of skull; but all of them are races of diminutive stature. This element, therefore, which has been referred to as important in individual comparisons, is no less necessary to be borne in view in determining such comparative results as those which distinguish the Peruvians from other American races. Certain races are unquestionably distinguished from others by difference of stature. Barrow determined the mean height of the Bushman, from measurements of a whole tribe, to be 4 ft. 3½ in. D'Orbigny, from nearly similar evidence, states that of the Patagonians to be 5 ft. 8 in. The internal capacity of the Peruvian skull, as derived from eighteen male and six female Quichua skulls in Dr. Davis's collection, is 70, while he states that of the Patagonian skull as 67 and of the Bushman as 65 ; but it is manifest that the latter figures, if taken without reference to relative stature, furnish a very partial index of the comparative volume of brain.

Professor Goodsir, as already noted, held that symmetry of brain has more to do with the higher faculties than mere bulk. In the case of the Peruvians the systematic distortion of the skull precludes the application of this test. But in the small Hindoo skull the fine proportions have been repeatedly noted. Dr. Davis, in describing one of a Hindoo of unmixed blood, born in Sumatra, says : " His pretty, diminutive skull is singularly contrasted with those of the races by whom, alive, he was surrounded ;"* and he adds : " The great agreement of the elegant skulls of Hindoos in their types and proportions, although not in dimensions, with those of European races, has afforded some support to that wide-spread and learned illusion, ' the Indo-European hypothesis.' The Hindoo skulls are generally beautiful models of form in miniature."

Mr. Alfred R. Wallace, in his " Malay Archipelago," discusses the value of cranial measurements for ethnological purposes; and, employing those furnished by Dr. J. B. Davis in his " Thesaurus Craniorum " as a " means of determining whether the forms and dimensions of the crania of the eastern races would in any way support or refute his classification of them," he finally selected as the best tests for his purpose—1. The capacity of the cranium ; 2. The proportion of the width to the length taken as 100 ; 3. The proportion of the height to the length taken as 100. But here again, unfortunately, the systematic distortion of the Peruvian skulls limits us to the first of those tests. There are, indeed, the eleven normal Peruvian crania selected as such from the numerous Ancon skulls brought by Professor Agassiz from Peru. But those are stated by Professor Wyman to be on an average less by six inches than the ordinary skull. Some partial results embodied in the following table admit of comparison with those based on the more ample data of Table IX. Dr. Lucae, in his " Zur Organischen Formenlehre," already referred to, gives the cranial capacity of single skulls of different races, selected as examples of each. In these, as in others already referred to, the capacity was determined with peas ; and the results—assumed to be given in Prussian ounces,—are dealt with here, as in the skulls of Heinse and Bünger. The experiments carried on for the purpose of

* "Thesaurus Craniorum," p. 148.

testing the process fully confirmed the results stated by Professor
Wyman as to the differences in apparent cubical capacity according
to the material employed.   Taking a sound Huron Indian skull, a
mean internal capacity of 1490 grms. was obtained by repeatedly
gauging it with peas, and of 1439·5 with rice.   The position of the
Negro, heading the list, serves to show the exceptional nature of the
evidence; though this is rather due to the inferiority of other examples, such as the Chinese and Greenlander, than to its greatly
exceeding the Negro mean.   In the first column the unzen, as Prussian ounces, are rendered in grammes.   The second column gives the
nearer approximation to the true specific gravity, according to the
standard referred to, based on a series of experiments undertaken for
me by Professor Croft, and assuming 82·5 grms. of peas to occupy
the space of 100 grms. of water.   The third and fourth columns
represent the estimated brain-weight, after the requisite deductions,
on the basis of s.g. of brain as 1·0408.

TABLE VIII.
COMPARATIVE CAPACITY OF RACES : LUCAE.

	Internal Capacity. Grms.	I. C. Corrected. Grms.	Brain-weight. Grms.	Brain-weight. Oz. Av.
Negro	1169·28	1424·12	1281·71	45·2
Chinese	1081·58	1364·48	1228·04	43·4
Nubian	1041·24	1313·54	1182·19	41·7
Floris	1033·93	1304·38	1173·94	41·4
Papuan	1030·42	1299·95	1169·96	41·3
Greenlander	1023·12	1290·74	1161·67	41·0
Javanese	995·06	1254·54	1129·91	39·8

In the following table the examples are derived from Dr. J. B.
Davis's tables, with the exception of the Peruvians.   For these I
have availed myself of Dr. Jeffreys Wyman's careful observations
on the large collection in the Peabody Museum, the results of which
confirm Dr. Morton's earlier data.   One further fact, however, may
be noted as a result of my own study of Peruvian crania, amply confirmed by the published observations of others, viz., that while the
Peruvian head unquestionably ranks among those of the microcephalous races, the range of variation among the Coast tribes appears
to be less than that even of the Australian.   Of this there is good
evidence, based on the comparison of several hundred crania.   But

exceptional examples of unusually large skulls may be looked for in
all races; and a few of such abnormal Peruvian or other skulls
would modify the mean capacities and weights in the following
table. Nevertheless the average results, as a whole, are probably a
close approximation to the truth:

TABLE IX.

COMPARATIVE CEREBRAL CAPACITY OF RACES.

RACE.	NUMBER.	CAPACITY. CUB. INCHES.	BRAIN-WEIGHT. Oz. Av.
European........................	299	92·3	47·12
English.........................	21	93·1	47·50
Asiatic.........................	124	87·1	44·44
Chinese.........................	25	92·1	47·00
Hindoos.........................	35	82·5	42·11
Negroes.........................	16	86·4	44·08
Negro Tribes....................	69	85·2	43·47
American Indians................	52	87·5	44·64
Mexicans........................	25	81·7	41·74
Peruvians.......................	56	75·0	38·25
Esquimaux.......................	13	91·2	46·56
Oceanic.........................	210	89·4	45·63
Javans..........................	30	87·5	44·64
Australians.....................	24	81·1	41·38

Looking for some definite results from the various data here produced, the deductions to which they seem to point may be thus
stated. While Professor Wyman justly remarks that the relative
capacity of the skull, and consequently of the encephalon, is to be
considered as an anatomical and not as a physiological characteristic:
relative largeness of the brain is nevertheless one of the most distin-
guishing attributes of man. Ample cerebral development is the
general accompaniment of intellectual capacity, alike in individuals
and races; and microcephaly, when it passes below well defined
limits, is no longer compatible with rational intelligence; though it
amply suffices for the requirements of the highest anthropomorpha.
Wagner thus definitely refers the special characteristics which separate man from the irrational creation to one member of the encephalon: "The relation of the lobes of the cerebrum to intelligence
may, perhaps, be expressed thus: there is a certain development of
the mass of the cerebrum, especially of the convolutions, requisite in

order to such a development of intelligence as divides man from other animals."

The important data accumulated by Morton, Meigs, Davis, Tiedemann, Pruner Bey, Broca, and others, by the process of gauging the skulls of different races, proceeds on the assumption of brain of a uniform density. But it seems by no means improbable that certain marked distinctions in races may be traceable to the very fact of a prevailing difference in the specific gravity of the brain, or of certain of its constituent portions; to the greater or less complexity of its convolutions; and to the relative characteristics of the two hemispheres. Moreover, it may be that some of those sources of difference in races may not lie wholly out of our reach, or even beyond our control. The diversity of food, for example, of the Peruvians and of the American Indian hunter-tribes was little less than that which distinguishes the Esquimaux from the Hindoo, or the nomad Tartar from the Chinese. The remarkable cerebral capacity characteristic of the Oceanic races is the accompaniment of well defined peculiarities in food, climate, and other physical conditions; and Australia is even more distinct in its physical specialties than in its variety of race.

Looking then to the unwonted persistency of the Peruvian cranium within such narrow limits, so far at least as the physical characteristics of the predominant population of Peru are illustrated by means of the great coast cemeteries; and to the striking discrepancy between the volume of brain and the intellectual activity of the race : I am led to the conclusion that, in the remarkable exceptional characteristics thus established by the study of this class of Peruvian crania, we have as marked an indication of a distinctive race-character as anything hitherto noted in anthropology.

www.ingramcontent.com/pod-product-compliance
Lightning Source LLC
Chambersburg PA
CBHW021635270326
41931CB00008B/1041